STUDENT UNIT GUIDE

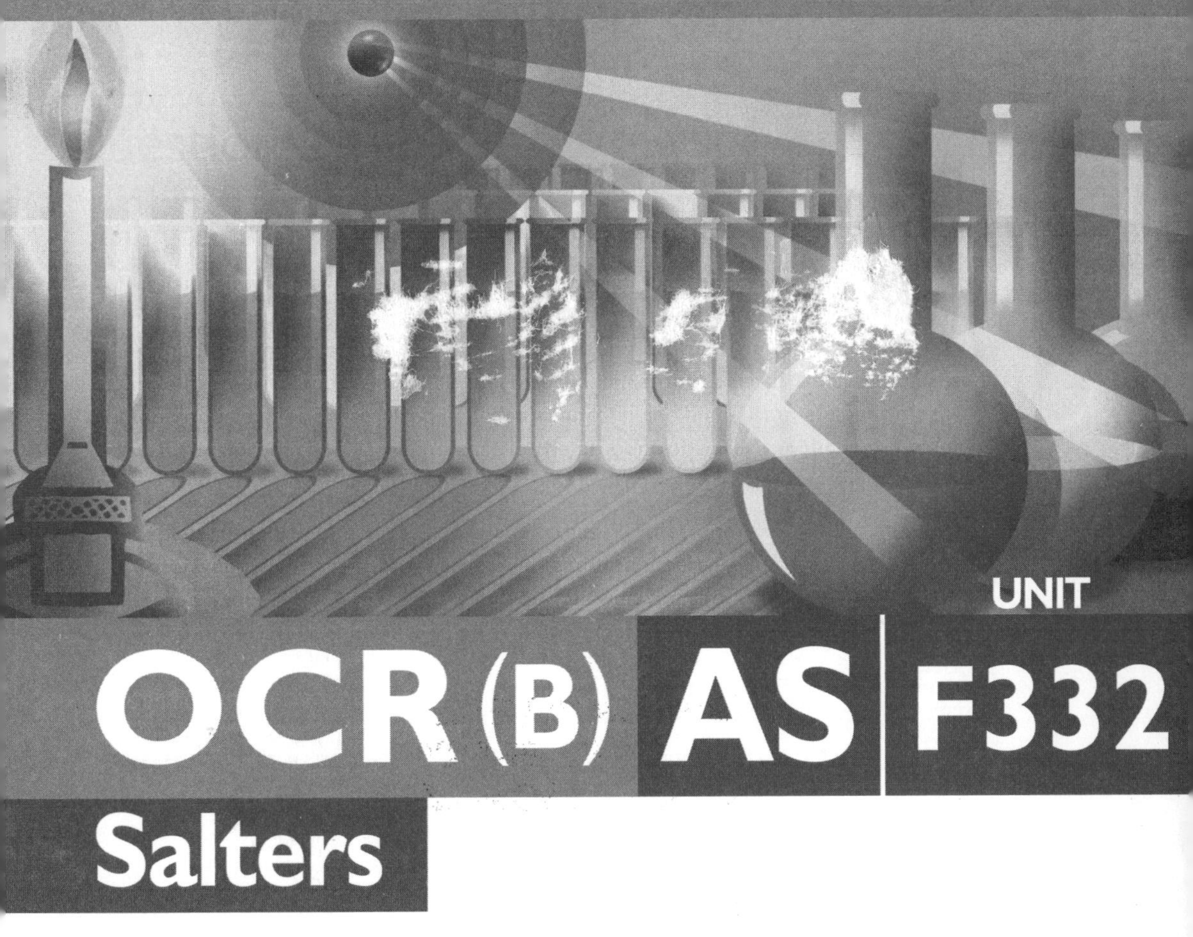

UNIT

OCR (B) AS F332

Salters

Chemistry

Chemistry of Natural Resources

Ashley Wheway

Philip Allan Updates, an imprint of Hodder Education, part of Hachette Livre UK, Market Place, Deddington, Oxfordshire OX15 0SE

Orders
Bookpoint Ltd, 130 Milton Park, Abingdon, Oxfordshire OX14 4SB
tel: 01235 827720
fax: 01235 400454
e-mail: uk.orders@bookpoint.co.uk

Lines are open 9.00 a.m.–5.00 p.m., Monday to Saturday, with a 24-hour message answering service. You can also order through the Philip Allan Updates website: www.philipallan.co.uk

ISBN 978-0-340-94822-4

First printed 2008
Impression number 5 4 3 2 1
Year 2013 2012 2011 2010 2009 2008

This guide has been written specifically to support students preparing for the OCR AS Chemistry (Salters) Unit F332 examination. The content has been neither approved nor endorsed by OCR and remains the sole responsibility of the author.

Typeset by Fakenham Photosetting Ltd, Fakenham, Norfolk
Printed by MPG Books, Bodmin

Hachette Livre UK's policy is to use papers that are natural, renewable and recyclable products and made from wood grown in sustainable forests. The logging and manufacturing processes are expected to conform to the environmental regulations of the country of origin.

P01298

Contents

Introduction

■ ■ ■

Content Guidance

■ ■ ■

Questions and Answers

Introduction

About this guide

This guide is designed to help you prepare for the second OCR (Salters) AS Chemistry Specification B unit examination, which examines the content of **AS Unit F332: Chemistry of Natural Resources**. The unit covers the following three teaching modules:

- Elements from the sea
- The atmosphere
- The polymer revolution

The aim of this guide is to provide you with a clear understanding of the requirements of the unit and to advise you on how best to meet them.

It is not intended as a last-minute crammer, and neither does it contain all the material that you need to be aware of. However, it does cover the facts and main 'chemical ideas' that you are required to know. During your studies, you will acquire much more information, which may help to improve your extended-writing responses.

The guide is divided into the following sections:

- This **Introduction**, which provides guidance on study, revision and examination techniques, showing you how to prepare for the unit examination.
- **Content Guidance**, which provides a summary of the facts and chemical ideas introduced and developed in the teaching modules of Unit F332.
- **Questions and Answers**, in which you will find questions in the same style as in the unit examination. Examiner's comments follow these answers. Careful consideration of the answers and comments will improve your:
  - understanding of the chemistry involved
  - examination technique
  - score in the examination

To understand chemistry, you have to be able to make links between chemical ideas. The Salters course gradually develops concepts through the different storylines. This approach leads to the continuous revision of chemical ideas. The first AS module consolidates and develops ideas from GCSE, as well as introducing new concepts and principles. It is important to make sure that your understanding of the GCSE and the ideas in Unit F331: Chemistry for Life are secure, before developing them further. A brief reminder of this prior knowledge is outlined at the start of each topic in the Content Guidance section.

The specification

The specification is an essential component of your study materials throughout the course. It is much more than a document to tell teachers what they should be doing and students what they should be learning. It outlines the format of the modules and contains details of the chemical knowledge and the depth of understanding you need to achieve to succeed. There is advice about a number of issues, such as:

- the correct ways of writing formulae
- a description of the 'How science works' topic is developed from GCSE
- guidelines for success in the practical assessments

If your teacher is not able to provide you with a full copy of the specification, it can be downloaded from the web at **www.ocr.org.uk**.

Note that OCR Chemistry B (Salters) has its own study materials: the *Storylines* and *Chemical Ideas* books and worksheets. In revising, you need to distinguish between what the specification says you *must* know and the extra material contained in the Salters materials. Your own notes, taken in class and enhanced through personal study, should be restricted to the specification guidelines and should be your main resource for revision. The most important contexts from the three storylines of this module are covered in the Question and Answer section of this guide.

The unit examination

The unit examination is a structured question paper that lasts 1 hour 45 minutes and is worth 100 marks. It counts for 50% of the AS marks, or 25% of the A-level marks. Usually, there are five questions. Question 5 of the paper is based on an 'advanced notice article' and is allocated 20 marks.

Your teacher will give you the article at some suitable time before the test. You should read carefully the instructions that come with the article. One of these tells you that, when answering the question, you are expected to apply your knowledge and understanding of the work covered in Unit F332.

Carefully read through the article itself, ensuring that you understand what it is saying. You may discuss it with other students and your teacher, and you may use books or the internet to find out more about the facts, ideas and issues raised by the article. Make some notes that you can look at the night before you take the unit.

- You should know the meanings of chemical words or phrases that are used in the article but not defined there.
- You should be able to expand on some of the chemical theory given in the article. For example, if it says 'a polymer may be made from X' but does not give the formula of the polymer, make sure that you can provide the formula.
- You should think about possible calculations or equations that are relevant and use them to expand the theory.
- You should be able to explain or interpret any theory, arguments or graphs presented in the article. For example, if it says 'carbon dioxide is a greenhouse gas', you should be able to explain how it acts in this way.

- You should be able to understand statements from the article and apply them to new situations.

Remember, however, that you will only be asked for meanings or theories that are in the *specification* for Unit F332 (i.e. in this book). Ask your teacher for advice if you are not sure.

Assessment objectives

The three assessment objectives (AOs) in this unit are described below.

AO1: knowledge and understanding

This means that you can be asked to:

- recognise and recall specific facts: e.g. 'What colour is an aqueous solution of bromine?' or 'What is the name of the functional group circled in the formula above?'
- recall laws, principles and concepts and be able to understand how to apply them to examples given in the specification: e.g. 'State Le Chatelier's principle. Use it to explain the effect of increasing the carbon dioxide concentration in the air on the acidity of sea water.'
- know the meaning of specified terminology and be able to use the words correctly when referring to the examples studied: e.g. 'Give the formula of the nucleophile in the reaction between bromobutane and aqueous sodium hydroxide.'
- select relevant information: e.g. 'Which of the following statements about electrode potentials is false?'
- organise information: e.g. 'The following statements refer to reactions which take place in the atmosphere. Put them in the order which explains how CFCs cause the destruction of ozone.'
- be able to communicate information in a variety of forms: e.g. 'Draw the skeletal formula of chlorobutane', 'Use the enthalpy diagram for the decomposition of hydrogen peroxide to comment on the enthalpy change for the reaction' or 'Using a table, compare electrophilic addition and nucleophilic substitution reactions.'

AO2: application of knowledge and understanding

With regard to this objective, you can be asked to analyse and evaluate scientific knowledge and processes (the term 'processes' includes collecting evidence, explaining, theorising, modelling, validating and planning to test an idea). For example, you should be able to explain which factor — bond enthalpy or bond polarisation, is more important in determining the rate of hydrolysis of halogenoalkanes.

AO2 also includes carrying out calculations. The calculations mentioned in the unit's specification involve:

- concentrations of solutions
- titration results
- determining the oxidation states of atoms in compounds and ions
- the equation $\Delta E = h\nu$

- composition by volume measured in percentage concentrations and in parts per million (ppm)

Calculations are an integral part of all chemistry at AS and A2. Therefore, some types of calculation met in Unit F331 could well be tested here. These include using the mole concept to determine masses of substances and volumes of gases, calculating empirical and molecular formulae and working out percentage composition.

You may also be asked to apply scientific knowledge and processes to unfamiliar situations, including those related to issues: e.g. you should be able to look at a flow diagram or equations for an industrial reaction sequence and be able to describe and explain the possible environmental issues involved. Given appropriate data for gases, you should be able to comment on whether they might lead to global warming, ozone depletion or acid rain.

You could be asked to assess the validity, reliability and credibility of scientific information. For example, 'Are Ziegler–Natta catalysts true catalysts? Are they not really initiators?' or 'In the 1980s, both the British and Americans collected data on ozone concentrations above Antarctica. In 1984 the British scientists discovered a 'hole' in the ozone layer. Initially, why did the Americans' data not support this discovery?'

AO3: how science works

This is *not* to be confused with the topic 'How science works', which is part of all the A-level units and is developed further from your GCSE studies.

For this section, you can be asked to:

- describe ethical, safe and skilful practical techniques and processes, selecting appropriate qualitative and quantitative methods: e.g. 'How would you use a colorimeter to measure the concentration of a MnO_4^-(aq) solution?' or 'Describe how you would use your titration results to calculate the amount of Fe^{2+} in a sample of "iron" tablets.'
- analyse, interpret, explain and evaluate the methodology, results and impact of given experimental data and investigative activities: e.g. 'Use the students' results to determine a value for the amount of Fe^{2+} in a sample of "iron" tablets. Explain why the method used leads to a lower value than that given on the label of the bottle.'

In the F332 examination paper, 40% of the marks must be of AO1 type, 46% of AO2 and 14% of AO3.

Command terms

The following command terms are often used in Salters examination questions. You must be clear about what they are asking you to do.

Describe

Tell the examiner about ... . This could be about a reaction, the properties of a

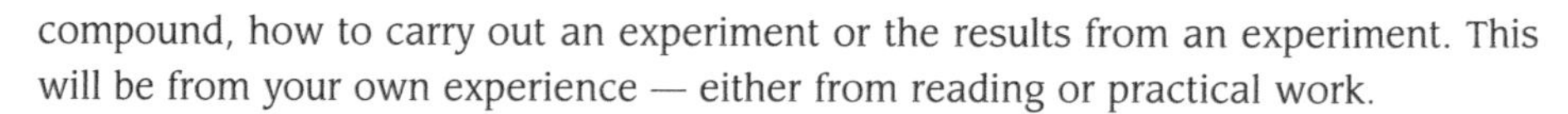

compound, how to carry out an experiment or the results from an experiment. This will be from your own experience — either from reading or practical work.

Explain
Use chemical ideas to say why and how things happen.

Suggest
Give a *possible* example, technique or reason.

Using...
You must make sure that whatever the examiner wants you to use is referred to in your answer.

Classify
Assign examples to a particular group or category.

Complete
This usually refers to a table, diagram, sentence or formula.

Draw
This used to apply to drawing chemical apparatus as two-dimensional sections. Increasing use of computers has made this largely a forgotten art. However, you should be able to recognise apparatus, spot errors and suggest possible improvements to such diagrams.

'Draw' also applies to constructing structural formulae. Make sure that you know how to draw full formulae correctly for the examples in this unit.

Preparing for the unit examination

Preparation begins when you start the AS course. The end result, the unit examination, is designed not only to show the examiner what you know and understand, but to show that you can organise information and communicate facts and ideas clearly and logically.

Reading

I give my students a weekly planner which tells them what each lesson covers and indicates the necessary prior reading from *Storylines* and *Chemical Ideas*. This reading enables the lesson time to be used effectively. Students will have noted points that need further clarification and sometimes areas that do not make any sense. As we discuss each topic, the students make connections quickly and develop an understanding more easily.

It is not easy to fit reading into your programme, but it provides a big bonus for those prepared to make the effort. You should make sure that you know what is coming next. With the Salters course, there is no excuse for not finding this out.

Note-taking

The use of loose-leaf paper is unsatisfactory. Pages can be lost for a variety of reasons, or can be mixed up. Hard-backed notebooks are preferable, with one side for in class and the adjacent side for follow-up work and revision comments. You should scan-read your lesson notes on the day that they are made, to check for errors. You will forget the details of the lesson if you leave it any longer. Remember that these notes are for revision, so they should be easy on the eye, as well as concise, clear and accurate. You should include some information from *Storylines, Chemical Ideas,* the Salters worksheets and your own research. You will need your notes if you continue to A2. You should ask your teacher to check them.

Study groups

Studying in groups is an important way of learning. You could form a group that meets regularly to sort out ideas, read each other's notes, prepare practical work, go over errors made during tests and talk through some of the more difficult exercises.

Revision

When the time comes for examination revision, do:

- have a plan — 'revision bytes' are a good idea, i.e. small chunks of material easily assimilated in a short period of continuous activity
- revise regularly — if you work in fits and starts, you will have difficulty retaining information
- test yourself — write the essentials out on scrap paper
- meet your study group — test each other

Past papers

Past papers are an important resource. You could complete some under examination conditions, and use others as a resource during revision on specific topics. The examiners' mark schemes and reports are also important. The mark schemes give information about what is and what is not acceptable for each answer. Your teacher should have copies of these or you can obtain them from OCR. The examiners' reports give details of where students did well and also areas where students generally need to improve. These documents form a valuable student resource.

Storylines

In the 48 hours before the unit examination, you should read through the relevant *Storylines* and take note of the assignments. This will remind you of the scope of the three modules and bring the contexts to mind. In addition, the assignments often contain the types of question that appear on the papers.

During the unit examination

- Make sure you look at the *Data Sheet*. It has a copy of the periodic table, which is an important resource. Make sure that you know how to use it effectively. The answers to some questions may require data about atomic numbers or relative atomic masses. The relative atomic masses are given to three significant figures in the periodic table, so do not try to use data from memory. It is also essential that you understand the information given in the table of characteristic infrared absorptions. Familiarise yourself with what peaks labelled M, M-S, S and S(broad) look like. The other data are for use at A2 only.
- Before you pick up your pen, *read the whole question paper*. This is important for two reasons. First, it allows you to select the question on which to make a start; this might not be the first question. Decide which question you will find the easiest and gain confidence by scoring heavily at the beginning. Second, while you are working on one question, the subconscious part of your brain may trigger helpful thinking for a different question.
- *Next read the question carefully.* Highlight or underline important command words and data given. As you cover each part, check the mark allocation. Have you made enough points? Have you shown your working in a calculation? These issues are covered in the Question and Answer section of the guide. By the time that the examination comes along, they should be second nature to you.
- Answers that require extended writing should be planned in rough first. You will have time to do this. If you do not have any paper, you will find space in your answer booklet. Examiners decide the appropriate amount of space for your response. If your writing is large, you may need a little more space, but if your writing is tiny, you will not need to fill it. Make sure that you are not rephrasing the question or using it as an introduction to your answer.
- Marks for quality of written communication are usually referred to as QWC marks. They are awarded for pieces of extended writing. They are not additional marks, but are awarded in the context of the answer to the question. It is, of course, important that your writing is legible.

  There are two types of QWC mark and they are indicated in the question by either of the following statements: 'In your answer you should make clear how your explanation links with the chemical theory' or 'In your answer, you should use appropriate technical terms, spelled correctly'.

Content Guidance

This section is a guide to the content of **AS Unit F332: Chemistry of Natural Resources**. The material is arranged under the following chemical topics:

- Atomic structure and the periodic table
- Bonding and structure
- Competition reactions
- Inorganic chemistry and the periodic table
- Calculations and moles
- Organic chemistry
- Isomerism
- Applications of organic chemistry
- Modern analytical techniques
- Equilibria
- Kinetics
- Catalysts

The essential facts are largely in tables or diagrams for easy access and revision. There is also guidance about specific chemical ideas and issues.

Do take note of the 'Prior knowledge' sections that head each topic. These remind you of the knowledge from GCSE and the modules of Unit F331: Chemistry for Life, which is assumed in this guide.

Answers to the questions in this section are on pages 50–52.

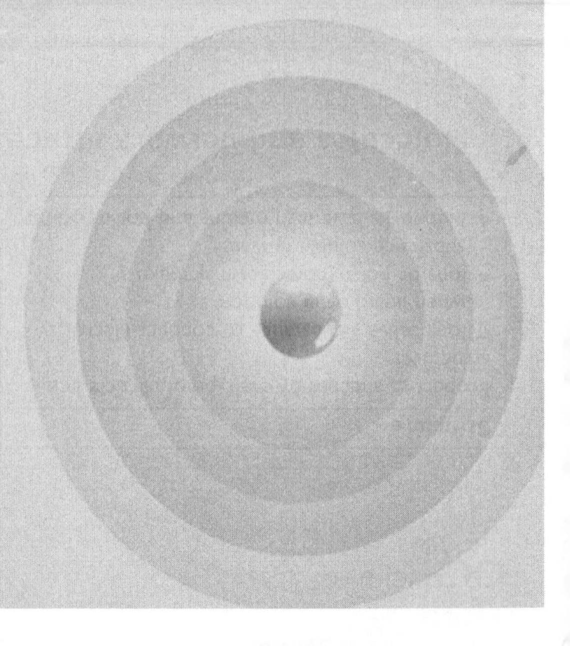

Atomic structure and the periodic table

Prior knowledge

It is assumed that you know:

- the electron structure of the first 20 elements in terms of shells and energy levels
- how to label shells, using *n*, the principal quantum number
- the meaning of the term 'ground state'

Sub-shells, orbitals and conventions for representing electrons

Atomic spectra of the elements show that electron shells are split into **sub-shells**. Sub-shells are labelled *s*, *p*, *d* and *f*. Different sub-shells can hold different numbers of electrons. The sub-shells themselves contain **atomic orbitals**.

At AS and A2, we associate orbitals with regions of space around the nucleus in which the electrons are moving. Although we can measure the energy that an electron has, we do not know its actual position. We can only know the *probability* of finding an electron at a particular position in an orbital.

Each orbital can hold up to two electrons. Electrons in orbitals can be thought of as spinning either clockwise or anticlockwise. If two electrons are present in an orbital, they have opposite spins.

The table below summarises the information about the sub-shells and orbitals that is needed to deduce the electron configuration of the elements from hydrogen to krypton.

Principal quantum number (*n*)	Sub-shell label	Number of atomic orbitals in sub-shell	Maximum number of electrons in sub-shell	Maximum number of electrons in shell
1	1*s*	1	2	2
2	2*s* 2*p*	1 3	2 6	8
3	3*s* 3*p* 3*d*	1 3 5	2 6 10	18
4	4*s* 4*p*	1 3	2 6	32

Krypton has eight electrons in its outer shell, but shell 4 can contain a maximum of 32 electrons if the 4*d* and 4*f* sub-shells are present.

Orbitals are often represented as labelled boxes and electrons as arrows. The diagram shows this for a 2*s* sub-shell containing two electrons and a 2*p* sub-shell with three orbitals, two of which are occupied by a single electron.

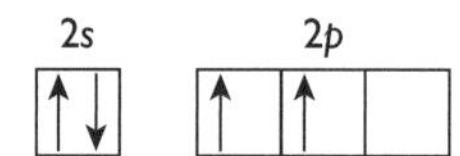

Electronic structure of the first 36 elements in the periodic table

The different atomic orbitals have different energy levels. The order of energy from lowest to highest for the atomic orbitals in the table on p. 13 is:

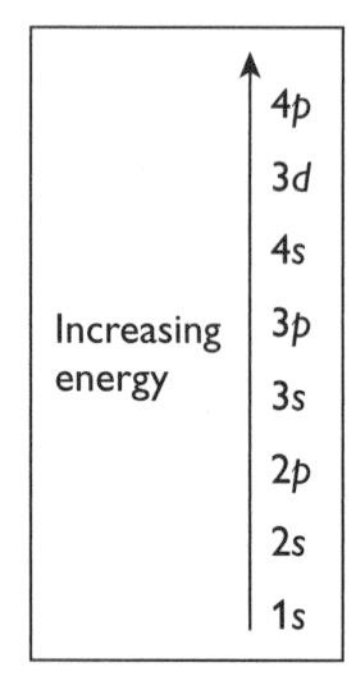

Remember that the 4*s*-orbital is slightly lower in energy than the set of 3*d*-orbitals.

The arrangement of the electrons of an atom in orbitals is called its **electron configuration**. Alternative terms for this are **electron arrangement** and **electronic structure**.

You can work out the ground-state electron configuration of any of the atoms of elements from H to Kr, using the following rules:

- Look up atomic number of the element (proton number).
- Add the same number of electrons to the orbitals in order of increasing energy.
- If there are two electrons in one orbital, they must have opposite spins.
- If there are several orbitals with the same energy (e.g. *p*- and *d*-orbitals), the electrons first go in singly with parallel spins.

The ground-state electron configurations for some of the elements are given in the table below. A shorthand notation is used to represent the electron configuration.

You should be able to write the electron configuration of the first 36 elements in the periodic table.

Element	Number of electrons	Electron configuration	Comment
H	1	$1s^1$	Only one electron, therefore easy
He	2	$1s^2$	Electrons have opposite spins
Li	3	$1s^2 2s^1$	Third electron must go up a level
B	5	$1s^2 2s^2 2p^1$	Fifth electron into new sub-shell
N	7	$1s^2 2s^2 2p^3$	All three electrons in *p*-orbitals have parallel spins
Ne	10	$1s^2 2s^2 2p^6$	2*p*-orbitals now filled
Na	11	$1s^2 2s^2 2p^6 3s^1$	Eleventh electron must go up a level
Ar	18	$1s^2 2s^2 2p^6 3s^2 3p^6$	3*p*-orbitals now filled
Ca	20	$1s^2 2s^2 2p^6 3s^2 3p^6 4s^2$	Electrons go into 4s before 3*d*
Sc	21	$1s^2 2s^2 2p^6 3s^2 3p^6 3d^1 4s^2$	Now 3*d*-orbitals fill before the 4*p*-orbitals
V	23	$1s^2 2s^2 2p^6 3s^2 3p^6 3d^3 4s^2$	All the *d*-electrons have parallel spins
Cr	24	$1s^2 2s^2 2p^6 3s^2 3p^6 3d^5 4s^1$	All *d*-electrons have parallel spins; $3d^5 4s^1$ (half-filled *d*-set) is more stable than $3d^4 4s^2$
Mn	25	$1s^2 2s^2 2p^6 3s^2 3p^6 3d^5 4s^2$	Back to normal
Ni	28	$1s^2 2s^2 2p^6 3s^2 3p^6 3d^8 4s^2$	Two *d*-orbitals are singly occupied and have parallel spins
Cu	29	$1s^2 2s^2 2p^6 3s^2 3p^6 3d^{10} 4s^1$	$3d^{10} 4s^1$ (filled *d*-set) is more stable than $3d^9 4s^2$
Ga	31	$1s^2 2s^2 2p^6 3s^2 3p^6 3d^{10} 4s^2 4p^1$	Back to normal again
Kr	36	$1s^2 2s^2 2p^6 3s^2 3p^6 3d^{10} 4s^2 4p^6$	Finished

Tip In writing the electron configuration of an atom, it is conventional to give any 3*d* electrons before the 4s etc.

Blocks in the periodic table

The elements in the periodic table are grouped into blocks. The first three of these are labelled the ***s*-block**, the ***p*-block** and the ***d*-block**. The outer orbital, which is being filled, gives its name to the block.

Question 1
In which period do the *f*-block elements first appear?

Bonding and structure

Prior knowledge

It is assumed that you know about:

- giant ionic and covalent structures (GCSE)
- shapes of molecules
- ionic, metallic, covalent and dative covalent bonding
- simple description of intermolecular bonding (GCSE)
- alkanes and alcohols
- enthalpy measurements

Relationship between structure and bonding

You need to be able to:

- identify whether particles in a substance are **atoms**, **molecules** or **ions**
- distinguish clearly between the terms **structure** and **bonding**

Question 2

(a) Make a list of rules that will help you to decide what type of particle is present in a substance, given its name or formula.

(b) Use your list to decide what type of particles are present in:

- **potassium chloride, KCl**
- **phosphorus(III) fluoride, PF_3**
- **barium, Ba**
- **poly(propene)**
- **aqueous hydrogen bromide**

You should be able to understand and appreciate the comparisons made in the following table.

Structure	**Bonding**
The way that atoms are arranged in a substance	The type of 'electron glue' that holds the atoms together in a given structure
Can be determined *experimentally*	A *theoretical* concept using models to explain how and why atoms form ions and molecules
Types of structure: • Ionic lattice (e.g. NaCl) • Metallic lattice (e.g. Na, Mg, Al) • Giant atomic lattice or network (e.g. diamond and silicon(IV) oxide) • Molecules (e.g. H_2O, Br_2, C_2H_6) • Polymers or macromolecules (e.g. poly(ethene))	Types of bonding: • Ionic • Metallic • Covalent • Dative covalent

Structure	Bonding
Properties affected by the type of particle in a structure: • Ions cause a substance to conduct electricity when liquid or in solution • Ions or polar molecules may make a substance soluble in water • Networks have high melting and boiling points	Properties affected by the strength of bonds: • Weak covalent bonds break easily (e.g. peroxides decompose readily, giving off oxygen) • Very strong bonds make molecules unreactive (e.g. nitrogen gas)

Molecules and network structures

Network is another name for 'giant'. You should be able to sketch the network structures of diamond and silicon(IV) oxide, showing how the atoms are joined together and the approximate bond angles.

In addition, for CO_2 and SiO_2, you must be able to:

- compare their properties
- describe their structures
- discuss their intramolecular bonding
- appreciate the role that intermolecular bonds play in deciding the properties of CO_2
- explain their properties in term of their structures and bonding

	Carbon dioxide	Silicon(IV) oxide
Formula	CO_2	SiO_2
Properties	Gas at room temperature; soluble in water	Solid with high melting point; insoluble in water
Structure	Molecular	Network (giant)
Intramolecular bonding	Covalent	Covalent
Intermolecular bonding	Very weak	Strong covalent bonds throughout the network

Sodium chloride, diamond and silicon(IV) oxide

The structures and properties of sodium chloride (NaCl), diamond and silicon(IV) dioxide (SiO_2) are summarised in the table below.

	NaCl	Diamond	SiO_2
Structure	Each ion is surrounded by six positively charged ions — this forms a cubic ionic lattice	Each carbon atom is covalently bonded, tetrahedrally, to four others — this forms a giant lattice (network)	Each silicon atom is covalently bonded, tetrahedrally, to four oxygen atoms — this forms a giant lattice with O–Si–O linkages (network)

	NaCl	Diamond	SiO_2
Hardness	Brittle — close, oppositely charged ions repel	Very hard — strong C–C bonds	Hard — strong Si–O bonds
Melting point	High (801°C) — strong net attractive force between oppositely charged ions in lattice	Very high (3820°C) — very strong bonds between atoms	Very high (1610°C) — very strong bonds between atoms
Solubility in water	Soluble — water forms strong ion–dipole interactions with ions	Insoluble — water molecules cannot interact with carbon atoms to break down the structure	Insoluble — water molecules cannot interact with silicon and oxygen atoms to break down the structure
Electrical conductivity	Does not conduct electricity in the solid state — ions cannot move (it does, however, conduct, and decomposes in the liquid state)	Does not conduct electricity — no ions or free electrons present	Does not conduct electricity — no ions or free electrons present

Electronegativity, polar bonds and polar molecules

Electronegativity is the ability of an atom in a molecule to attract electrons. Electron pairs in bonds are not equally shared between the two atoms. If the value of the electronegativity difference between the two atoms is significant, then the bond is **polar**. This means that the more electronegative of the two atoms has a slight negative charge and the other a slight positive charge.

For example:

Partial charges $\delta+$ $\delta-$

C —— F

Electronegativity values 2.6 4.0

Remember that C–H bonds are considered to be non-polar.

There are two requirements for a molecule to be polar, i.e. to have a permanent dipole:

- It must have at least one polar bond.
- If there is more than one polar bond, the individual bond dipoles must not cancel each other out.

The second point depends upon the shape of the molecule. Consider carbon dioxide and sulfur dioxide molecules. Both have polar bonds, but look at their shapes:

$\delta-$ $\delta+\delta+$ $\delta-$

O=C=O

The dipoles cancel; the molecule is non-polar

$\delta+\delta+$

S

$\delta-$O O$\delta-$

The dipoles do not cancel; the molecule is polar

A sulfur dioxide molecule has a negative end and a positive end; a carbon dioxide molecule does not.

Intermolecular bonds

The table below summarises what you need to know about intermolecular bonds.

	Type of intermolecular bond		
	Instantaneous dipole–induced dipole	**Permanent dipole–permanent dipole**	**Hydrogen bonding**
Particles that it occurs between	All	Molecules with a permanent dipole	Suitable hydrogen atom on one molecule, small electronegative atom (F, O, N) on the other
Approximate strength	0.1–1.0 kJ mol^{-1}	1–10 kJ mol^{-1}	10–100 kJ mol^{-1}
Example	Propane	Ethanal	Ethanol
Reason	C–H bonds are considered to be non-polar	Permanent dipole due to the C=O bond H_3C, δ+ C=O δ–, H The hydrogen atoms are not suitable for hydrogen bonding	Suitable hydrogen atom and an oxygen atom C_2H_5–O δ– ··· δ+ H–O–C_2H_5
M_r of example	44	44	46
Boiling point of example	–42°C	21°C	78°C

Do not use the term 'van der Waals forces' when you mean instantaneous dipole–induced dipole bonds. Van der Waals forces are a collection of forces and include interactions between permanent dipoles and induced dipoles.

A 'suitable hydrogen' for hydrogen bonding is usually one that is attached to an oxygen, nitrogen or fluorine atom. Sometimes a hydrogen atom attached to a carbon atom that also has several halogen atoms attached is suitable. The electronegative atoms cause the hydrogen atom to have a slight positive charge (δ+).

Question 3

Trichloromethane and propanone molecules form hydrogen bonds with each other. Draw a diagram to show one hydrogen bond between the two molecules.

Competition reactions

Prior knowledge

You are expected to know about:

- formulae of ions (GCSE)
- oxidation and reduction (GCSE)

Redox in terms of electron transfer

Redox reactions are **competition** reactions, in which particles compete for electrons. You should know and understand the following *three* ways of describing a redox reaction.

Oxidation	Reduction
A substance gaining oxygen	A substance losing oxygen
A substance losing electrons	A substance gaining electrons
The oxidation state of an atom increasing	The oxidation state of an atom decreasing

Remember:

- the mnemonic '**oilrig**' applies to electron transfer, *not* to changes in oxidation state (**o**xidation **i**s **l**oss of electrons, **r**eduction **i**s **g**ain)
- an **oxidising agent** *gains* electrons and is *reduced*
- a **reducing agent** *loses* electrons and is *oxidised*

Oxidation states

The oxidation state of an atom is best considered as simply a number (with a sign) that can be calculated for each atom in a molecule or ion, using a set of rules. It is a measure of the control that an atom has over its electrons, and can help us to decide whether oxidation and reduction are occurring in a given reaction. These rules are also important in naming molecules and ions and in helping to construct redox equations. These last two uses are studied at A2.

Assigning oxidation states to atoms

Some of the rules for assigning oxidation states are summarised in the table below.

Atom	Oxidation state
Uncombined elements	0
Group 1	+1
Group 2	+2
Aluminium	+3
Fluorine	–1

Atom	Oxidation state
Oxygen	–2
Hydrogen	+1
Chlorine, bromine or iodine	–1

The following exceptions should be noted:

- Oxygen has an oxidation state of –1 in peroxides (these contain an O–O bond) and +2 in OF_2.
- Hydrogen has an oxidation state of –1 in ionic hydrides (H^-).
- Chlorine, bromine and iodine have higher oxidation states than –1 when combined with oxygen or a more reactive halogen (e.g. in ClF_3 the oxidation state of chlorine is +3).

You also need to remember that:

- in a *compound*, the sum of the oxidation states of all the atoms in the formula is *zero*
- in an *ion*, the sum of the oxidation states of all the atoms in the formula equals the *charge on the ion*

Tip Students often make mistakes in calculating states where there is more than one atom of the same kind present. For instance, in calculating the oxidation state of the chlorine atom in Cl_2O, they look at the formula and remember correctly that the sum has to be zero and that oxygen is always –2. They arrive at the incorrect answer of +2, whereas for each chlorine atom the answer is +1.

Question 4

Calculate the oxidation state of:

- **sulfur in $Na_2S_2O_3$**
- **nitrogen in NO_3^-**
- **phosphorus in P_2O_3**
- **nitrogen in NH_4^+**

Tip The term 'oxidation number' is also used at AS level to mean oxidation state. Exam papers will use the term 'oxidation state'. You will not lose credit if you use 'oxidation number'.

Hydration of ions in solution

Some ionic solids dissolve in water because the polar water molecules are able to interact with the ions. The weak dipoles on water molecules are attracted to the ions, releasing energy. This energy is sufficient to overcome the attraction between the oppositely charged ions in the solid compound. Ions in solution are said to be **hydrated**.

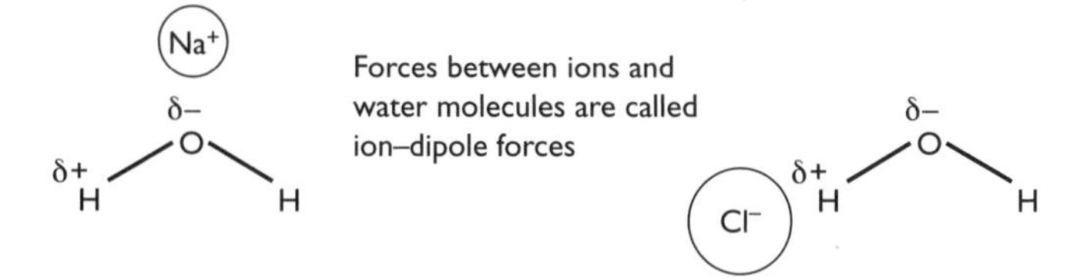

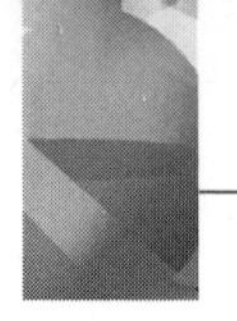

The average number of water molecules surrounding an ion depends on the size and charge of the ion.

Question 5
Explain why non-hydrated chloride ions are larger than non-hydrated sodium ions.

Inorganic chemistry and the periodic table

Prior knowledge

You are expected to know about the trends in group 7 from GCSE.

Some trends in the periodic table

Charges on atomic ions

Not only are you expected to recall how the charge on an atomic ion varies with its position in the periodic table as shown below, you must also be able to explain this relationship in terms of ionisation enthalpies.

Group	1	2	3	4	5	6	7
Charge on ion	1+	2+	3+	4–	3–	2–	1–

Ionisation enthalpies

You need to be able to define the term ionisation enthalpy and write equations for the successive ionisations of an element.

Question 6
(a) What is meant by the term 'first ionisation enthalpy of magnesium'?
(b) Write an equation for the second ionisation enthalpy of magnesium.

In order to explain how the first ionisation enthalpy varies with atomic number and the charge on an atomic ion, you need to understand the *factors* that affect the *size* of ionisation enthalpies, which are as follows:

- the distance of the outer shell from the nucleus
- the number of protons in the nucleus
- the 'shielding' of the outermost electron from the nuclear charge by the filled inner shells of electrons

Question 7

(a) Why does the first ionisation enthalpy of an element in group 2 decrease down the group?

(b) Why does the first ionisation enthalpy of an element generally increase across a period?

(c) Why does magnesium form ions with a charge of 2+ rather than 3+?

Precipitation reactions and ionic equations

It is important to remember that ions in solution behave *independently*. This means that, when a solution of ions takes part in a reaction, often only one type of ion is involved. Ions that do not take part in the reaction are called **spectator** ions. This makes it easier to understand what is happening in the reaction and to write the correct ionic equation.

A **precipitation reaction** takes place when two solutions containing ions are mixed and a solid forms.

To write an ionic equation, follow these steps:

- **Step 1**: identify the solid that forms.
- **Step 2**: write its formula on the right-hand side of the equation, followed by the '(s)' state symbol.
- **Step 3**: write down the formulae of the ions that come together to form the solid, followed by the '(aq)' state symbols.
- **Step 4**: balance the equation.

Example

Observations

When aqueous silver(I) nitrate(V) is added to a potassium chloride solution, a white precipitate forms.

Writing the ionic equation

Remember that the positive ion from one solution reacts with the negative ion from the other to form a solid that is insoluble in water.

In this case, the precipitate is either silver(I) chloride or potassium nitrate(V). How can we decide? You are expected to recall this information.

In general, it is worth remembering two solubility rules:

(1) All group 1 compounds are soluble.
(2) All nitrates are soluble.

(These rules will enable you to identify the name of the precipitate in any such reaction that you are likely to meet at AS or A2.)

(1) The precipitate is silver(I) chloride.
(2) $\rightarrow AgCl(s)$
(3) $Ag^{+}(aq) + Cl^{-}(aq) \rightarrow AgCl(s)$
(4) The equation is balanced.

You should be able to write ionic equations for the following reactions of aqueous halide ions with aqueous silver nitrate.

Positive ion in aqueous silver nitrate	Negative ion in aqueous metal halide	Appearance of precipitate	Formula of precipitate
$Ag^+(aq)$	$Cl^-(aq)$	White	$AgCl(s)$
$Ag^+(aq)$	$Br^-(aq)$	Cream	$AgBr(s)$
$Ag^+(aq)$	$I^-(aq)$	Yellow	$AgI(s)$

Group 7: the halogens

Physical properties

Property	Chlorine	Bromine	Iodine
State at room temperature	Gas	Liquid (volatile)	Solid (volatile)
Appearance at room temperature	Greenish-yellow	Brown (orange vapour)	Dark grey, crystalline (purple vapour)
Appearance in water	Pale greenish-yellow	Orange	Pale yellow (brown in aqueous KI)
Appearance in organic solvents	Pale greenish-yellow	Orange	Purple

Displacement of halide ions by halogens

Halide ions (Cl^-, Br^- and I^-) are colourless in aqueous solution. Therefore, any colour change is due to one halogen molecule displacing another.

If you can remember the colours of the halogens when dissolved in water, and their relative reactivity, you will be able to predict and write an ionic equation for any one of these reactions. In these reactions, the *halogen is the oxidising agent* and the *halide ion is the reducing reagent.*

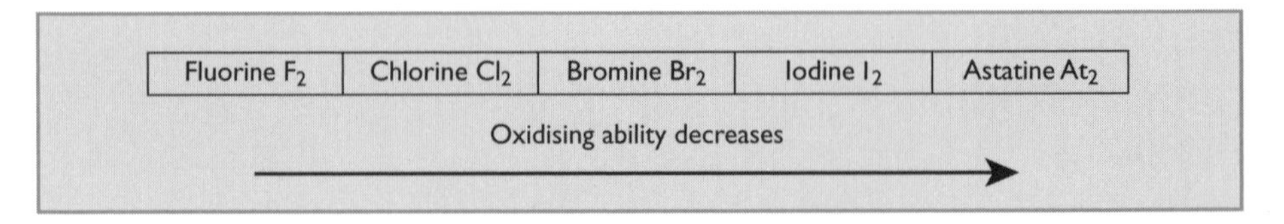

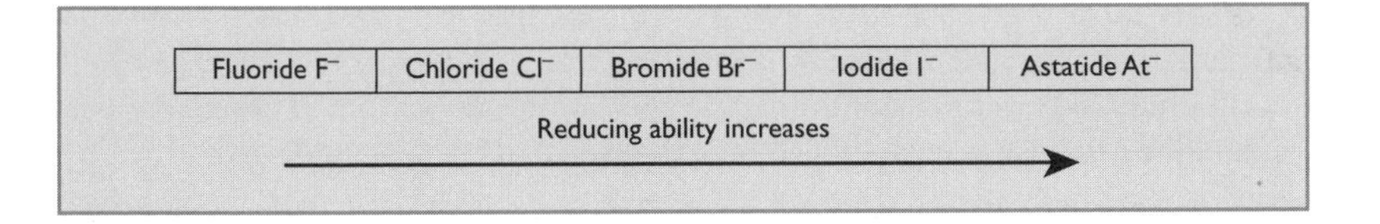

Question 8

(a) What will you see when aqueous chlorine is added to aqueous potassium iodide?

(b) Write the ionic equation for the reaction between aqueous chlorine and aqueous potassium iodide.

Calculations and moles

Prior knowledge

You are expected to know how to:

- use the mole concept to perform calculations involving masses of substances
- use the mole concept to perform calculations involving volumes of gases
- construct balanced chemical equations

Setting out calculations

You should set out your working so that examiners can give you credit for knowing how to tackle a given problem.

Whenever you use an 'equals sign', the left-hand side should contain at least one word or symbol that defines the quantity that you are attempting to calculate. The right-hand side should not be only numbers and mathematical signs — it should contain the appropriate units.

Concentration of solutions

Equation 1

$$\text{amount of substance (mol)} = \frac{\text{mass (g)}}{\text{molar mass (g mol}^{-1}\text{)}}$$

Equation 2

$$\text{concentration (mol dm}^{-3}\text{)} = \frac{\text{amount of substance (mol)}}{\text{volume (dm}^{-3}\text{)}}$$

At AS, the calculations are structured. This means that each calculation is split up into steps, with each step requiring a different type of calculation or deduction. All the data required are provided in the question or on the *Data Sheet.* Therefore you should look carefully at what you have been given.

Example

What mass of sodium hydroxide (NaOH) is present in 100 cm^3 of a 0.100 mol dm^{-3} solution?

(A_r: H, 1.00; O, 16.0; Na, 23.0)

In the calculation, start with the concentration of the standard solution (a solution whose concentration is known accurately), and use Equation 2 (p. 25) to find the amount of sodium hydroxide in the solution.

moles of NaOH = concentration × volume of solution

= 0.100 × (100/1000) mol

Now use Equation 1 (above) to calculate the mass of NaOH used.

M_r of NaOH = 23.0 + 16.0 + 1.0 = 40.0

mass of NaOH = 0.100 × (100/1000) × 40.0 g

= 0.400 g

Tip

- Most students tend to calculate relative formula mass (M_r), rather than molar mass. This is fine as long as you remember that M_r does not have units.
- Remember that the volume in a concentration expression has the units dm^3, not cm^3.
- Do not use the abbreviations 'ml' or 'cc'.
- Do not use your calculator until you arrive at the last step. This reduces the likelihood of making arithmetical errors.
- Make sure that you use the correct number of significant figures in your final answer.

Acid–base titration

You should be able to use a given balanced equation or information about ratios of reacting moles to determine volumes or concentrations of reacting solutions.

Example

25.00 cm^3 of a solution of limewater (calcium hydroxide solution) are titrated with 0.0300 mol dm^{-3} hydrochloric acid solution. The titre is 23.50 cm^3. Calculate the concentration of the calcium hydroxide in the limewater.

The equation for this reaction is:

$$Ca(OH)_2 + 2HCl \rightarrow CaCl_2 + 2H_2O$$

Start with the concentration of the standard solution and use Equation 2 (p. 25) to find out the amount of HCl in the titre.

moles of HCl in titre = concentration × volume of solution:

$$= 0.0300 \times (23.50/1000) \text{ mol}$$

moles of $Ca(OH)_2$ needed to neutralise the HCl titre:

$$= 0.5 \times 0.0300 \times (23.50/1000) \text{ mol}$$

concentration of $Ca(OH)_2 = 0.5 \times 0.0300 \times (23.50/1000) \times 1000/25.00 \text{ mol dm}^{-3}$

$$= \mathbf{0.0141 \text{ mol dm}^{-3}}$$

Tip

- You may find it helpful to write the data given, as well as what you are trying to find, under the appropriate substances in the equation. This makes it easier to see which substance is the standard solution. For the above example:

$Ca(OH)_2$	+	2HCl	→	$CaCl_2 + 2H_2O$
25.00 cm^3		23.50 cm^3		
? mol dm^{-3}		0.0300 mol dm^{-3}		

- In the second step, you need to refer to the equation for the correct ratio of acid to base.
- Wait until the last step to use your calculator. The 1000s always cancel — a useful check.

Volume composition

Two methods are used to measure the proportions of each gas present in the atmosphere:

- concentration by volume, expressed as a percentage
- parts per million (ppm), which is used for substances present in small concentrations

For example, carbon dioxide has a concentration in the atmosphere of 367 ppm. This means that in 10^6 parts measured by volume, 367 parts are carbon dioxide. In other words, there are 367 cm^3 of the gas in 10^6 cm^3 of the atmosphere.

Expressing this as a percentage by volume:

concentration of carbon dioxide = $(367 \times 100)/10^6 = 0.0367\%$

Prior knowledge

You are expected to know:

- about structural isomerism and molecular shapes
- the difference between aliphatic and aromatic compounds
- how to recognise the following homologous series: alkanes, cycloalkanes, alkenes, alcohols and ethers
- about the combustion of alkanes and alcohols
- that simple addition polymers are formed from alkene monomers and you should be able to draw the structure of the repeating unit of a polymer

Homologous series

The homologous series (for Units F331 and F332) that you need to be able to recognise are summarised in the table below. You must also be able to name **alkenes** and **halogenoalkanes**.

Homologous series	General formula	Functional group	Example
Alkanes	C_nH_{2n+2}	n/a	$CH_3CH_2CH_2CH_3$ Butane
Cycloalkanes	C_nH_{2n}	n/a	(hexagon) C_6H_{12} Cyclohexane
Alkenes	C_nH_{2n}	>C=C<	$H_3CHC{=}CHCH_3$ But-2-ene
Halogenoalkanes	$C_nH_{2n+1}X$, where X is a halogen atom	—X	$(CH_3)_3CBr$: H_3C—C(CH_3)(Br)—CH_3 2-bromo-2-methylpropane
Alcohols	$C_nH_{2n+1}OH$	—O—H	H_3C—C(H)(OH)—CH_3 Propan-2-ol

Homologous series	General formula	Functional group*	Example
Ethers	$C_nH_{2n+1}OC_mH_{2m+1}$	R—O—R	CH_3—O—C_2H_5 Methoxyethane
Aldehydes	$C_nH_{2n}O$	—C(=O)—H	CH_3CH_2CHO Propanal
Ketones	$C_nH_{2n}O$	—C(=O)—R	H_3C—C(=O)—CH_3 Propanone
Carboxylic acids	$C_nH_{2n+1}COOH$	—C(=O)—O—H	CH_3CH_2COOH Propanoic acid
*R = alkyl group			

Naming organic compounds

There are three steps to naming an organic compound. A halogenoalkane is used here as an example.

H_3C—CH(F)—CCl(F)—CH_3

Step 1: Identify the longest carbon chain and name it as the appropriate alkane.

H_3C—CH(F)—CCl(F)—CH_3 The alkane is butane

Step 2: Identify the substituent halogen groups and add them as a prefix, putting the groups in alphabetical order of the halogen. Use di, tri etc. to signify the number of similar substituent groups. Ignore these prefixes when arranging the groups alphabetically.

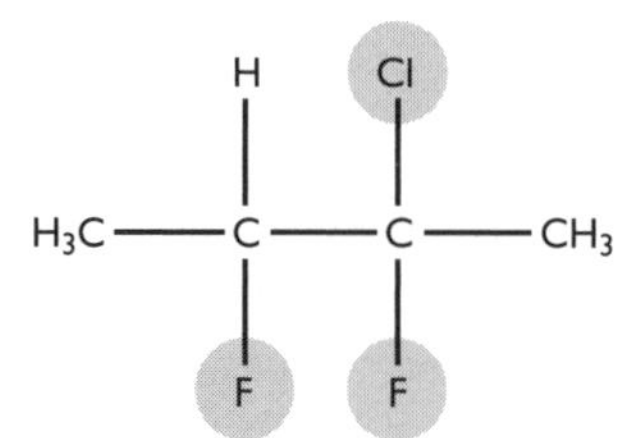

The prefix is chlorofluoro so the name is chlorodifluorobutane.

Step 3: Number the substituents from the end of the carbon chain to give the lowest number of substituent positions.

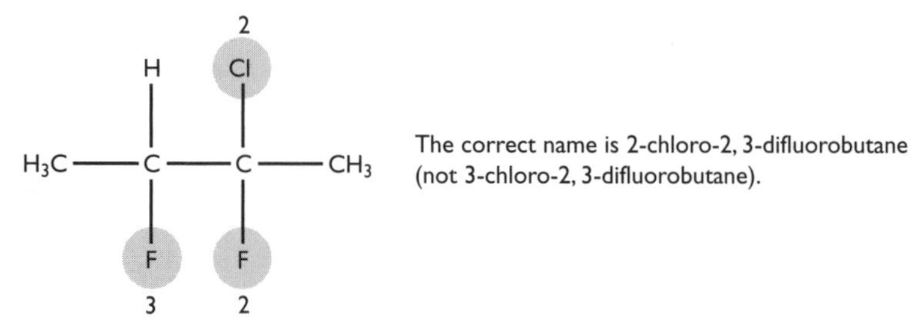

The correct name is 2-chloro-2, 3-difluorobutane (not 3-chloro-2, 3-difluorobutane).

Alcohols

Alcohols with only one hydroxyl group (OH) are called **monohydric** alcohols. Monohydric alcohols can be classified as **primary, secondary** or **tertiary alcohols**.

Type of alcohol	Position of hydroxyl group	Example
Primary	End of chain (attached to a carbon atom that has at least two hydrogen atoms attached)	Butan-1-ol $H_3CH_2CH_2C$—C(H)(H)—OH
Secondary	Middle of chain (attached to a carbon atom that has one hydrogen atom attached)	Butan-2-ol H_3C—C(H)(OH)—CH_2CH_3
Tertiary	Attached to a carbon atom that has no hydrogen atoms attached	2-methylpropan-2-ol H_3C—C(CH_3)(OH)—CH_3

Reactions to learn

You need to know the reagents and conditions for a number of reactions. Examiners have a variety of ways of asking questions about these. Make sure that you are familiar with the facts about each reaction and how it is used. Some are used to make other substances (synthesis). Others are used to help find out about the functional group present in a substance. Examples are given in the Question and Answer section of this guide.

Compound	Reagent	Conditions	Products
Alkane	Chlorine, Cl_2	Ultraviolet light	Chloroalkanes and HCl
Alkene	Bromine, Br_2	Room temperature	Dibromoalkane
Alkene	Aqueous HBr	Room temperature	Bromoalkane
Alkene	Hydrogen, H_2	Platinum catalyst at room temperature (or nickel catalyst with heat (150°C) and pressure)	Alkane
Alkene	Water, H_2O	Steam and phosphoric acid catalyst with heat (300°C) and pressure (60 atm) (or water and concentrated sulfuric acid)	Alcohol
Halogenoalkane (RX)	Aqueous sodium hydroxide	Heat under reflux	Alcohol and a sodium halide (NaX)
Halogenoalkane (RX)	Concentrated aqueous ammonia, NH_3	Heat in a sealed tube	Amine and HX (react further to produce a salt)
Primary alcohol	Acidified dichromate(VI) solution ($Cr_2O_7^{2-}$)	Heat and distil off product	Aldehyde
		Heat under reflux	Carboxylic acid
Secondary alcohol	Acidified dichromate(VI) solution ($Cr_2O_7^{2-}$)	Heat	Ketone
Alcohol	Al_2O_3	Heat	Alkene
	Conc. H_2SO_4	Heat under reflux	

The oxidation of alcohols depends on the position of the hydroxyl group. These reactions are summarised in the table below.

	Primary	Secondary	Tertiary
Example of alcohol	H_3C—C(H)(OH)—H Ethanol	H_3C—C(CH_3)(OH)—H Propan-2-ol	H_3C—C(CH_3)(OH)—CH_3 2-methylpropan-2-ol
Oxidation product(s)	H_3C—C(=O)—H Further oxidation occurs, producing ethanoic acid H_3C—C(=O)—OH	H_3C—C(=O)—CH_3 Propanone	No reaction
Colour change of acidified $K_2Cr_2O_7$ (oxidising agent)	Orange to green	Orange to green	No change

Synthesis of halogenoalkanes from alcohols

There are three stages in preparing an organic compound by a synthetic route involving a single reaction step. These are:

- reaction
- separation
- purification

These steps are summarised in the flow diagram below, using the preparation of 1-chlorobutane from butan-1-ol as an example.

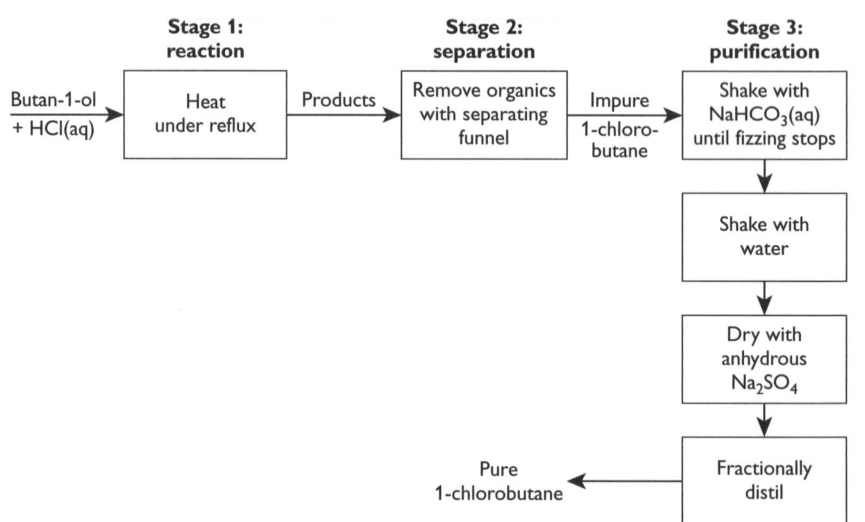

Question 9

(a) Why is it possible to remove the 'organics' with a separating funnel?

(b) Why is the organic layer shaken with first $NaHCO_3$(aq) and then water?

(c) Suggest a substance that is removed by distilling the dried chlorobutane.

Reaction mechanisms

The best possible mechanism of a reaction for a chemist would be a sort of molecular 'movie'. It would be a record of everything that happens from when the reacting particles approach each other to when the product molecules are formed. Unfortunately, this amount of information is not available. However, we can make computer animations using information gained about crucial points (like single frames of a movie) in the progress of the reaction. These 'still images' are often all we have on which to base our 'reaction mechanisms'.

In order to understand and write effective answers to questions on this topic, you need to be familiar with bond fission and the types of attacking reagent.

Bond fission means bond breaking. There are two ways in which a covalent bond can break. These are compared in the table below, using the C–Cl bond in chloromethane as an example.

	Homolytic fission (homolysis)	**Heterolytic fission (heterolysis)**
What happens to the bonding electron pair?	One electron goes to each atom H | H—C—Cl | H ↓ H | H—C• + •Cl | H	Both electrons go to the more electronegative atom H | H—C—Cl | H ↓ H | H—C^+ + :Cl^- | H
Species (atoms or group of atoms) formed	Radicals	Ions; if the resulting positive charge is on a C atom, the ion is called a carbocation
How does bond polarity favour the process?	Usually non-polar (e.g. C–H)	Usually polar (e.g. C–Cl)
Conditions favouring the process	Light (sometimes visible but usually ultraviolet); gas phase; non-polar solvents	Polar solvents

The species that attack carbon atoms in organic molecules are:

- **radicals** — atoms, molecules or even ions that have at least one unpaired electron. **Biradicals** have two unpaired electrons. The most common example of a biradical is dioxygen (O_2). This is contrary to what dot-and-cross diagrams would have us believe. In the dot-and-cross diagram for dioxygen, all electrons are paired, as shown below.

  However, a more complex model (not covered at A-level) indicates that there is an unpaired electron on each oxygen atom. This model explains the pale-blue colour of liquid oxygen, its magnetic properties and its reactivity.
- **nucleophiles** — molecules or negative ions with a lone pair of electrons that can attack an electron-deficient carbon (positively charged) to form a covalent bond. Examples include H_2O, OH^- and NH_3.
- **electrophiles** — molecules or positive ions that can attack an electron-rich carbon (region of high electron density) accepting a pair of electrons to form a covalent bond. Examples include Br_2 and H^+.

Chain reactions

Chain reactions are extremely fast and often require visible or ultraviolet light (e.g. the reaction between chlorine and methane). These are known as **radical chain reactions**. A radical chain mechanism has three stages.

- **Inititiation:** $Cl_2 \rightarrow 2Cl\bullet$
- **Propagation:** $CH_4 + Cl\bullet \rightarrow \bullet CH_3 + HCl$
  $\bullet CH_3 + Cl_2 \rightarrow CH_3Cl + Cl\bullet$
- **Termination:** $Cl\bullet + Cl\bullet \rightarrow Cl_2$
  $\bullet CH_3 + Cl\bullet \rightarrow CH_3Cl$
  $\bullet CH_3 + \bullet CH_3 \rightarrow C_2H_6$

It is important to notice that the first propagation step involves the chlorine radical (Cl•) abstracting a hydrogen atom from a methane molecule, rather than attacking a carbon atom to form chloromethane and a hydrogen radical. Hydrogen abstraction reactions are common in the troposphere because radicals are plentiful.

Question 10

Why are replacements for CFCs designed to contain C–H bonds?

Types of reaction

Most organic reactions can be classified as redox, acid–base, **substitution**, **addition** or **elimination**. Substitution, addition and elimination involve one of three groups of attacking species — **radicals**, **electrophiles** or **nucleophiles**. The reaction of chloromethane with aqueous sodium hydroxide is an example of nucleophilic substitution.

Tip When asked to name the mechanism for a given reaction, you need to choose one word from each of the two categories given in bold above.

The mechanisms of the reactions covered at AS are summarised in the table below.

Type	Organic reactant	Reagent	Type of attacking species	Organic product
Substitution	Alkane	Cl_2 with UV light	Radical	Halogenoalkanes
Substitution	Halogenoalkane	H_2O* or OH^-	Nucleophile	Alcohols
Substitution	Halogenoalkane	NH_3	Nucleophile	Amines
Substitution	Alcohol	X^- (halide)	Nucleophile	Halogenoalkanes
Addition	Alkene	Br_2	Electrophile	Dibromoalkanes
Addition	Alkene	HBr	Electrophile	Bromoalkane
Addition	Alkene	H_2O	Electrophile	Alcohols
Elimination	Alcohol	H_2SO_4	Not required	Alkenes

**Nucleophilic substitution reaction by water is also called hydrolysis.*

Question 11

Describe the difference between a hydrolysis reaction and a hydration reaction.

Describing mechanisms

Formulae, curly arrows and a few words are all you need to fully describe a mechanism. You need to be able to describe the mechanisms for **nucleophilic substitution**, **electrophilic addition** and **radical chain reactions**.

Remember, curly arrows show the movement of electrons. An electron pair is indicated by a full arrow (↷). A single electron is represented by a half-arrow (⌒).

Examples of nucleophilic substitution and electrophilic addition reactions are given below.

Nucleophilic substitution, e.g. bromomethane reacting with strong alkali

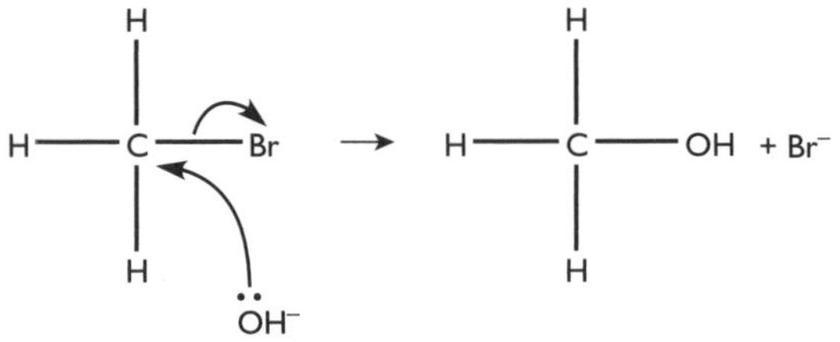

H, H, H, C, Br, OH⁻ → H, H, H, C, OH + Br^-

Electrophilic addition, e.g. bromine reacting with ethene

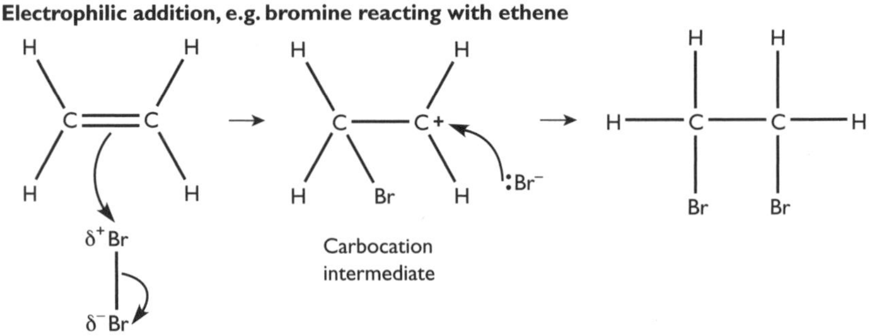

H, H, C=C, H, H, δ^+Br, δ^-Br → H, H, C, C+, H, Br, H, :Br^- → H, H, H, C, C, H, Br, Br

Carbocation intermediate

Question 12

Describe the mechanisms shown above for:

(a) nucleophilic substitution

(b) electrophilic addition

Which descriptive approach — labelled diagrams or written text — do you prefer?

Reactivity of halogenoalkanes

The **rate of hydrolysis** of halogenoalkanes is determined by reacting them with silver nitrate dissolved in aqueous ethanol. The purpose of the ethanol is to enable the halogenoalkane and aqueous solution to mix. The results for the reaction of various halogenobutanes at 60°C are given in the table below.

	1-chlorobutane	1-bromobutane	1-iodobutane
Bond broken in reaction	C–Cl	C–Br	C–I
Electronegativity difference between carbon and halogen	0.6	0.4	0.1
Bond enthalpy (kJ mol^{-1})	346	290	228
Time for precipitate to appear	No precipitate after 30 minutes	About 15 minutes	About 2 minutes
Colour of precipitate	—	Cream	Yellow
Formula of precipitate	—	AgBr	AgI

These results indicate that the relative rates of reaction of halogenoalkanes depend on the *bond strength* of the carbon–halogen bond.

Tip Beware — in examinations, students often incorrectly choose the electronegativity factor in their explanations of halogenoalkane reactivity.

You should be able to write equations for the hydrolysis and precipitation reactions occurring. If you have difficulty with precipitation reactions, then refer to pp. 23–24.

Question 13
Why does the rate of hydrolysis of halogenoalkanes depend on the strength of the carbon–halogen bond? Suggest a reason why chloroalkanes hydrolyse faster than fluoroalkanes.

Isomerism

E–Z isomerism

Two conditions are needed for a compound to show this type of isomerism:

- There must be *restricted rotation* at room temperature about a bond. Single bonds freely rotate, but double bonds cannot rotate without an input of energy or the use of a catalyst to break one of the bonds. The commonest bond which gives rise to geometric isomers is the C=C bond.
- There must be *two groups that can be arranged differently in space* by placing them either on opposite sides (*E*-isomer) or on the same side (*Z*-isomer) of the double bond.

Question 14
Explain why 1,2-dibromocyclohexane forms *E*- and *Z*-isomers.

In alkenes which have the same two groups on both sides of the double bond (e.g. 1,2-dibromoethene), the *E*- isomer is often referred to as *'trans'* and the *Z*-isomer as *'cis'*.

There are three alkenes that are structural isomers with the molecular formula C_4H_8:

$H_2C{=}CHCH_2CH_3$
But-1-ene

$H_3CHC{=}CHCH_3$
But-2-ene

$H_2C{=}C(CH_3)_2$
2-methylprop-1-ene

There are two *E–Z* isomers of but-2-ene

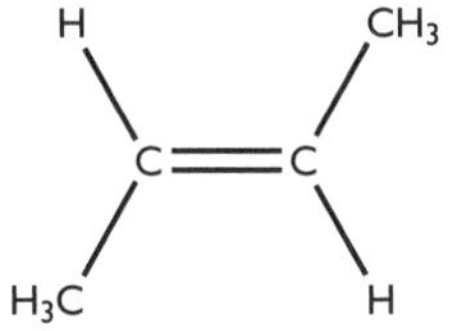

H, CH_3, C=C, H_3C, H
E (or *trans*) but-2-ene

and

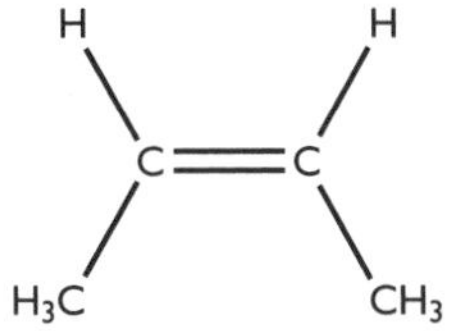

H, H, C=C, H_3C, CH_3
Z (or *cis*) but-2-ene

Drawing the shapes of isomers

In examinations, marks are often lost for poor drawings of structural or skeletal formulae and for labelling bond angles incorrectly.

Common errors include:

- the wrong number of bonds given around the carbon, nitrogen and oxygen atoms, or OH groups drawn as being bonded to a carbon by the hydrogen atom
- a hydroxyl group not shown as O–H when a full structural formula is required
- an incorrect number of carbon atoms in a chain or group when converting skeletal to structural formulae or vice versa
- guessing bond angles in organic molecules, often stating an impossible 90°

You are expected to be able to be able to draw structural formulae for molecules containing single and double bonds.

By applying the concept of areas of electron repulsion (see Unit F331) you should be able to give the approximate bond angles as 109° or 120°.

The diagrams below indicate how you should draw structures, using wedges and dotted lines to represent the three-dimensional shape where appropriate.

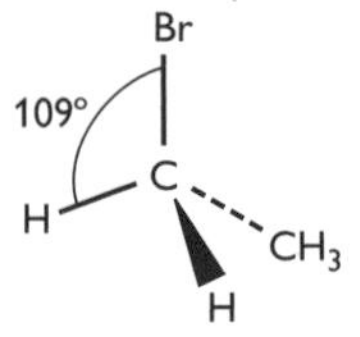

Br
109°
C
H
CH_3
H

Compound **A**

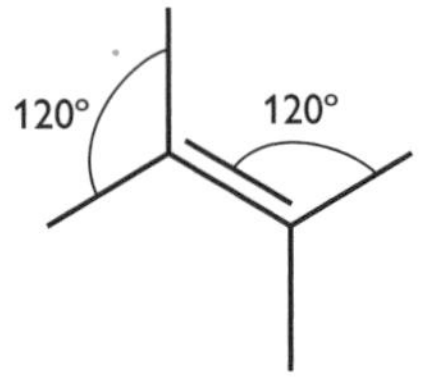

120°
120°

Compound **B**

> **Question 15**
> **Give the systematic names of compounds A and B.**

Applications of organic chemistry

Chlorofluorocarbons (CFCs)

You need to:

- know their uses and relate them to their properties
- understand their role in ozone depletion
- be able to explain why chemists thought they were safe
- be able to discuss the advantages and disadvantages of replacement compounds

CFCs contain C–F and C–Cl bonds, both of which are strong. Therefore, CFCs are unreactive in the troposphere. When CFCs eventually reach the stratosphere, ultraviolet radiation breaks the C–Cl bond homolytically.

Question 16
In a CFC molecule in the stratosphere, C–Cl rather than C–F bond breakage occurs. Explain why.

The uses and advantages of CFCs are shown in the table below.

Use of CFCs	Advantages
Refrigerants for food refrigerators and air-conditioning Propellants for aerosols Blowing agents for making expanding plastics Cleaning solvents for dry cleaning and for cleaning electronic circuitry	• Unreactive • Non-toxic • Not flammable • Correct volatility (e.g. for refrigerants: low enough to vaporise but high enough to liquefy under pressure) • Dissolve grease easily

CFCs and some replacement compounds compared

The advantages and disadvantages of CFCs and some types of replacement compounds are summarised in the following table. ODP stands for ozone depletion potential and is a measure of the ability of a compound to destroy ozone in the stratosphere. ODP is measured relative to CCl_3F (CFC-11), which is assigned a value of 1.0.

Group of compounds	Advantages	Disadvantages
CFCs	Advantages listed above	High ODP (1.0) Are greenhouse gases
HCFCs (hydrochlorofluorocarbons)	Low ODP (about 0.1)	Are greenhouse gases More expensive Some are flammable
HFCs (hydrofluorocarbons)	ODP = 0	Are greenhouse gases Much more expensive Some are flammable
Alkanes	ODP = 0 Cheap	Are greenhouse gases Flammable Difficult to find compounds with the appropriate properties

Polymers

Poly(propene) is a polymer which has been designed. Many polymers have been discovered by accident. These include low-density poly(ethene), poly(tetrafluoroethene) (PTFE), conducting and light-emitting polymers, and bakelite.

Only **addition polymers** are studied at AS. Addition polymers are formed from alkene **monomers**. If two or more different monomers are polymerised together, then the polymer is called a **copolymer**.

$$*n\ \mathrm{HHC{=}CHX} \xrightarrow{\text{addition reaction}} \left[-\mathrm{CH_2}-\mathrm{CHX}-\right]_n$$

Monomer — Repeating unit

*n is a very large number and X is an H atom or substituent group.

Common substituent groups are shown in the table below.

Alkene	Substituent group	Name of polymer
$H_2C{=}C(H)CH_3$	Methyl	Poly(propene)
$H_2C{=}C(H)C_6H_5$	Phenyl	Poly(phenylethene) or styrene
$H_2C{=}C(H)Cl$	Chloro	Poly(chloroethene) or polyvinyl chloride (PVC)

Plastics contain other materials as well as the polymer. These may be pigments, lubricants, antioxidants or materials to increase the strength of the plastic, such as glass wool. Plastics can be moulded to a defined shape.

Plastics are either **thermoplastics** or **thermosets**.

Thermoplastics

Also known as **thermosoftening plastics**, these contain polymers that soften on heating, but become stiff and solid-like again on cooling. This means that they can be remoulded. The polymer chains are able to move relative to each other.

Thermosets

These are also known as **thermosetting plastics**. When these are first moulded, the heat causes 'crosslinks' to form between polymer chains. The resulting 'plastic' is really a covalently bonded network of atoms and is hard and rigid. On heating, bonds eventually will break and the material 'chars' as carbon is formed.

Modern analytical techniques

Prior knowledge

You are expected to know about:

- common gases in the atmosphere (GCSE)
- interaction of radiation with matter

Quantisation of energy

Quantisation of energy means that particles can only have certain fixed amounts of energy. When energy changes, it can only change by fixed amounts. Some important facts about the energy associated with electrons and molecules are summarised in the table below.

Energy	Spectrum corresponding to the spacing between the energy levels	Energy change caused by absorption of radiation (J)
Electrons becoming excited	Visible and ultraviolet	1×10^{-19} to 1×10^{-16}
Bonds vibrating	Infrared	1×10^{-20} to 1×10^{-19}
Molecules rotating	Microwave	1×10^{-22} to 1×10^{-20}
Molecules translating (moving around)	The spacing is so tiny that translational energy is treated as continuous	

You should know that the regions in the electromagnetic spectrum listed below increase with frequency.

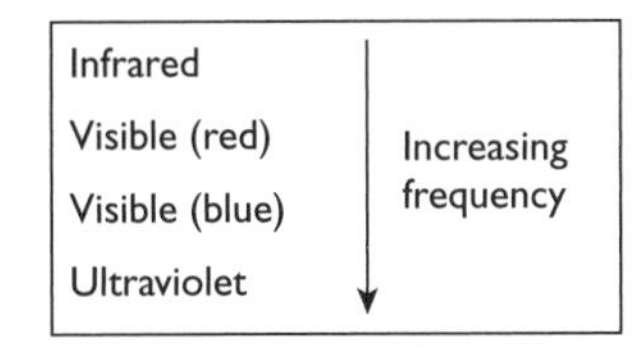

Question 17
Explain why it takes more energy for a C–F bond to vibrate than a C–Cl bond.

What can happen when molecules absorb ultraviolet radiation?

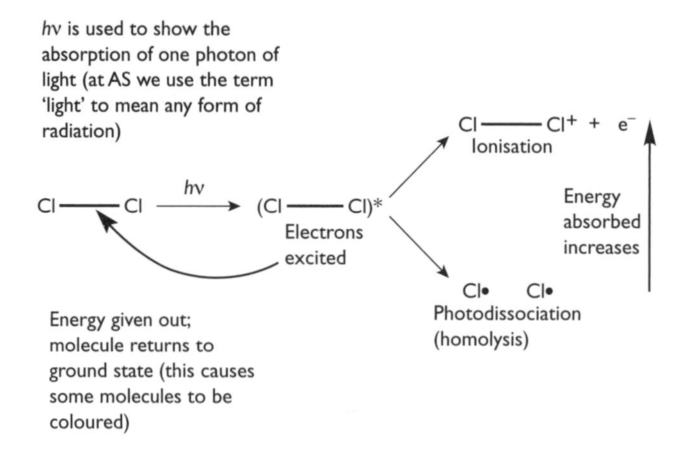

Ozone as a sunscreen

The formation and natural destruction of ozone occurs in the stratosphere. The formation of ozone takes place in two steps:

- Step 1 is called **photodissociation**:
  $O_2 + h\nu$ (ultraviolet) $\rightarrow O + O$
- Step 2 is the reaction between oxygen atoms (produced by photodissociation) and oxygen molecules, to produce ozone:

  $O + O_2 \rightarrow O_3$

The natural destruction of ozone is also a photodissociation reaction:

$O_3 + h\nu$ (ultraviolet) $\rightarrow O_2 + O$

This last equation is responsible for the screening effect of ozone.

Breaking a bond in ozone

You should be able to use the expression:

$E = h\nu$

to calculate the **frequency** needed to break a bond in an ozone molecule. You will be given its **bond enthalpy** and the **Planck constant** (h).

Example
The energy needed to break the O–O bonds in ozone is 392 kJ mol^{-1}.

The energy required to break one O–O bond is $392/6.02 \times 10^{23}$ kJ.

frequency (ν) = E/h

where $h = 6.63 \times 10^{-34}$ J Hz^{-1}.

$\nu = (392 \times 1000/6.02 \times 10^{23})/6.63 \times 10^{-34}$ Hz

$= 9.82 \times 10^{14}$ Hz

Tip This is where attention to units pays dividends. Bond enthalpies are given in kJ mol^{-1}, but the units of h are J Hz^{-1}, hence the factor of 1000.

The greenhouse effect

A greenhouse gas is one that absorbs infrared radiation in the troposphere. Neither visible nor ultraviolet radiation is absorbed. Some common gases in the atmosphere that absorb infrared radiation are given in the table below.

Name	Formula	Concentration in the troposphere (ppm)	Greenhouse factor*
Nitrogen	N_2	7.8×10^5	Negligible
Oxygen	O_2	2.1×10^5	Negligible
Argon	Ar	1.0×10^4	Negligible
Water vapour	H_2O	c. 1×10^4 (variable)	0.1
Carbon dioxide	CO_2	3.7×10^2	1
Methane	CH_4	1.8	20
CFC-11	$CClF_3$	2.6×10^{-4}	3800
CFC-12	CCl_2F_2	5.3×10^{-4}	8100

**The greenhouse factor compares the greenhouse effect for a gas molecule with carbon dioxide, which is assigned a value of 1.*

Question 18

Using the data in the table, explain why carbon dioxide makes the largest contribution to the total greenhouse effect.

If you are asked to explain the greenhouse effect, base your answer on the following outline:

- Most radiation, usually visible and ultraviolet (high frequency) from the Sun, that reaches the troposphere, is *absorbed* by the Earth.
- The Earth's surface warms up and *emits infrared radiation*.
- Some of this emitted infrared radiation is *absorbed* by greenhouse gases.
- The greenhouse gas molecules gain *vibrational* energy (or bonds in the molecules vibrate faster).

- The average *kinetic energy* of the gas in the troposphere increases, leading to an increase in the average *temperature* of the atmosphere.
- Greenhouse gas molecules *re-emit* some of the absorbed infrared radiation, some of which also *heats up* the Earth.

Tip The words in italics are key words which an examiner will expect to see used correctly by good candidates.

The evidence from relating increased concentrations of greenhouse gases to the increase in the Earth's surface temperature during the past 50 years is irrefutable. This temperature increase can only be explained by using a model that takes into account the increased greenhouse gas emissions. What is not certain is the effect that global warming will have on climate change.

Infrared spectroscopy

Absorption of infrared radiation causes the bonds in molecules to vibrate faster. Bonds can vibrate by 'stretching' or 'bending'. Not all vibrations can absorb infrared radiation.

The frequencies of radiation absorbed by a molecule:

- can give information about the *type of bonds* in the molecule and therefore the **functional groups** present
- are distinctive for a particular molecule, and can act as a 'fingerprint' *for identifying a compound* when compared with spectra in a large database.

You need to remember a few important features about an infrared spectrum:

- the so-called '**peaks**' are really 'troughs'
- the intensity of a 'peak' (how big it is) is labelled as **strong, medium-strong** or **medium**
- some peaks are **broad** due to hydrogen bonding
- the **frequency** of infrared radiation is measured in **cm^{-1}** and covers the range from 4000 cm^{-1} down to about 800 cm^{-1}
- the part of the spectrum below about 1500 cm^{-1} is often referred to as the **'finger-print' region**, because most of the absorptions ('peaks') are caused by complex vibrations of a particular molecule. You should not try to assign these absorptions to particular functional groups.

At AS, you only have to be able to identify — from an infrared spectrum — the presence of three functional groups. These are summarised in the table below.

Functional group	Bond responsible for absorption	Absorption frequency (cm^{-1})	Intensity of absorption
Hydroxyl	O–H O–H (hydrogen bonded)	3600–3640 3000–3600	Strong Strong (broad)
Carbonyl	C=O	c. 1700–1750	Strong
Carboxylic acid	O–H (hydrogen bonded) C=O	2500–3200	Medium (broad)

You have to know whether absorptions are medium or strong, sharp or broad. In the spectrum below, the C–H absorption is medium and the C=O strong; both are sharp. The position of C–H absorptions indicates whether the molecule is saturated or unsaturated (alkene or arene); see the table below.

Absorption frequency (cm^{-1})	Type of hydrocarbon
Above 3000	Unsaturated
Below 3000	Saturated

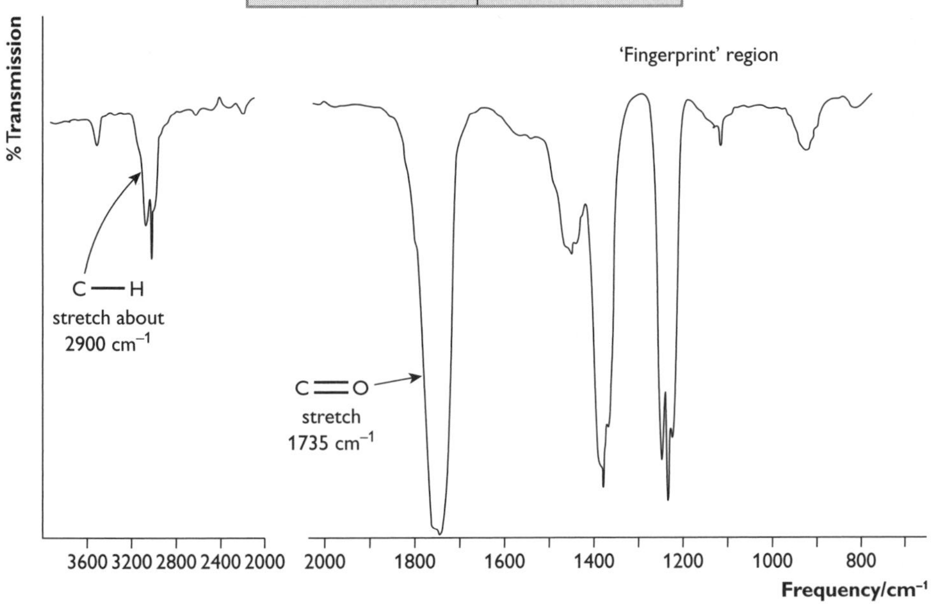

Part of the infrared spectrum of an alcohol is shown below. The spectrum shows that some O–H groups are hydrogen bonded while others are not. Note how hydrogen bonding causes broadening of the 'peak' and also a decrease in the frequency of the absorption.

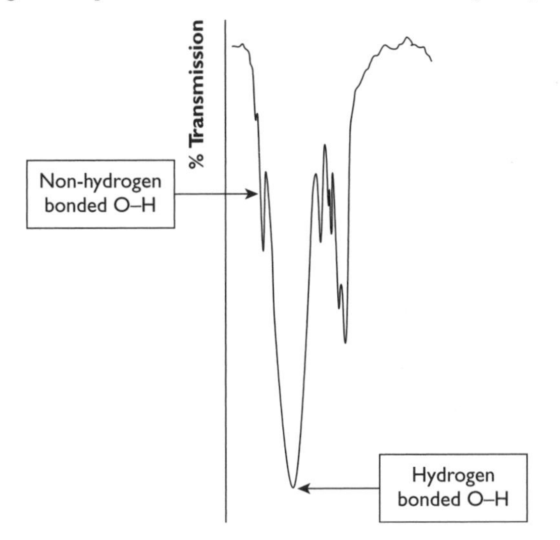

The carbonyl absorption frequency is not always exactly the same, but varies slightly with the environment of the group as shown in the table below. These values are given on the *Data Sheet*, and you should be able to apply the data, given a spectrum with a sufficiently detailed frequency scale or a list of specific frequencies.

Question 19

The infrared spectrum of compound C is shown below. There is only one functional group in compound C. Identify this group.

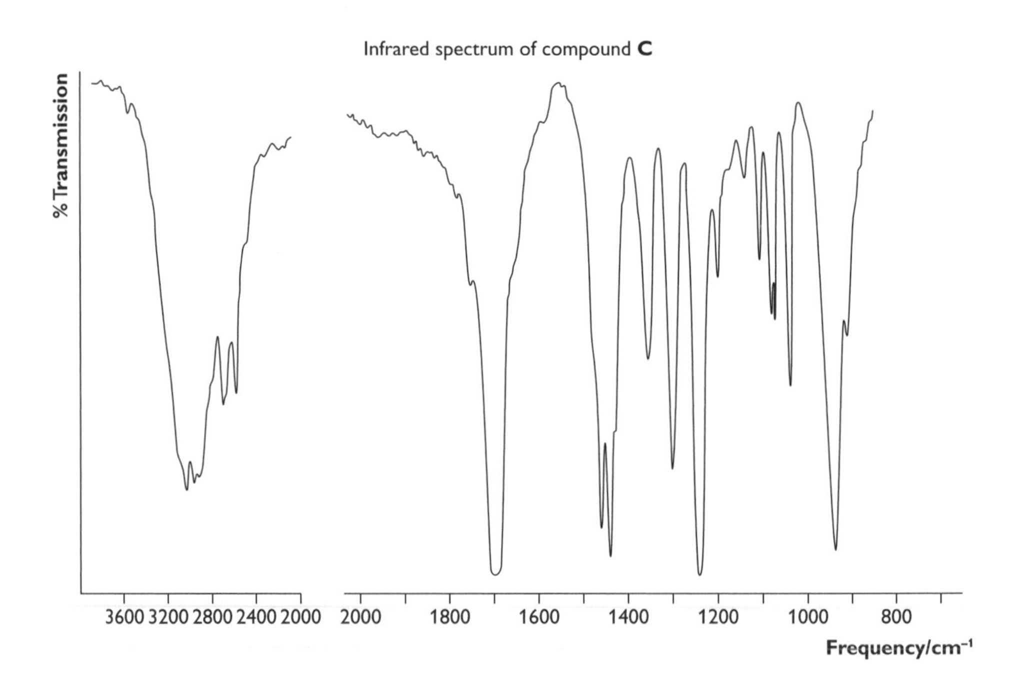

Type of carbonyl group	Absorption frequency (cm^{-1})	Intensity of absorption
Aldehyde	1720–1740	Strong
Ketone	1705–1725	Strong
Carboxylic acid	1700–1725	Strong
Ester	1735–1750	Strong
Amide	1630–1700	Medium

Equilibria

Prior knowledge

You are expected to know about reversible reactions.

Dynamic equilibrium

The concept of chemical equilibrium only applies to **closed** systems.

A closed system is one that cannot exchange mass with its surroundings. For a reaction in an open vessel to be a closed system, there must be no gaseous reactants or products. If gases are present, then the vessel must have a lid for chemical equilibrium to be achieved.

Chemical equilibrium involves **dynamic** processes. Nothing *appears* to be changing. However, reactant particles are colliding and forming product particles and product particles are also colliding, forming reactant particles. At equilibrium, the rates of these two processes are equal.

Using Le Chatelier's principle

Le Chatelier's principle states that, *for a system in equilibrium*, when a change is made in any of the conditions, the equilibrium position will move in the direction that counteracts the change.

The effects of changing conditions on the equilibrium position of some reactions are summarised in the table below.

Condition change	Effect on equilibrium position
Increasing the *concentration* of reactants	Moves to the right
Increasing the *concentration* of products	Moves to the left
Decreasing the *concentration* of reactants	Moves to the left
Decreasing the *concentration* of products	Moves to the right
Increasing the *pressure* for a reaction in which the products have fewer gas molecules than the reactants	Moves to the right
Decreasing the *pressure* for a reaction in which the products have fewer gas molecules than the reactants	Moves to the left
Increasing the *temperature* for a reaction in which the forward reaction is exothermic	Moves to the left
Decreasing the *temperature* for a reaction in which the forward reaction is endothermic	Moves to the right

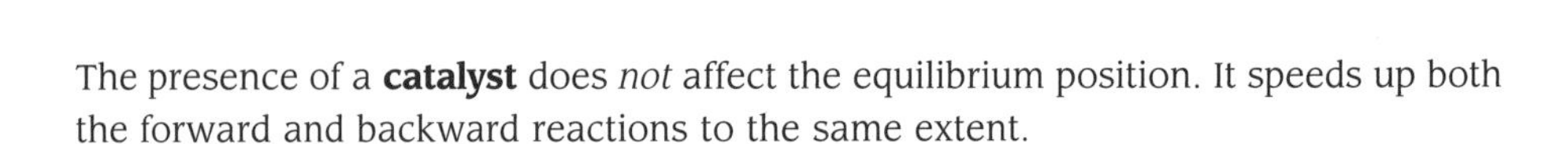

The presence of a **catalyst** does *not* affect the equilibrium position. It speeds up both the forward and backward reactions to the same extent.

Carbon dioxide in the atmosphere and the oceans

The atmosphere and the oceans are treated as a closed system. Gaseous carbon dioxide dissolves in water and reacts to form a weakly acidic solution. You should be able to write the equations for the equilibria occurring:

$CO_2(g) \rightleftharpoons CO_2(aq)$

$CO_2(aq) + H_2O(aq) \rightleftharpoons HCO_3^-(aq) + H^+(aq)$

Question 20

The solubility of CO_2 is greater at the bottom of the ocean than near the surface. Use Le Chatelier's principle to explain this.

Kinetics

Prior knowledge

You are expected to know about:

- the factors affecting the rate of a reaction (GCSE)
- simple collision theory (GCSE)
- heterogeneous catalysis
- enthalpy changes

Factors affecting reaction rates

The factors that are studied at AS are summarised in the table below.

Factor affecting rate of reaction	Explanation using collision theory
Concentration of reactants (in gases, the concentration is proportional to the pressure)	Reacting particles collide The collisions that have kinetic energy greater than or equal to the activation enthalpy lead to products An increase in concentration/pressure increases the total number of collisions; the number of collisions leading to products also increases
Temperature	An increase in temperature increases the average kinetic energy of the particles* A greater number of collisions will have energies greater than the activation enthalpy More collisions will be successful in forming products

Factor affecting rate of reaction	Explanation using collision theory
Presence of a catalyst	A catalyst provides an alternative reaction pathway that has a *lower* activation enthalpy A greater number of collisions will have energies greater than the activation enthalpy, so the rate is increased
Particle size of a solid reactant or catalyst	A decrease in particle size means an increase in surface area The number of collisions at the surface of the solid increases, so the reaction rate increases
Intensity of light or ultraviolet radiation in photochemical reactions	An increase in the intensity of light or ultraviolet radiation increases the number of bonds broken, forming more radicals, so the reaction rate increases

**There is also a slight increase in the collision frequency with increasing temperature, but the main effect on reaction rate is from the increase in the average kinetic energy of the particles.*

Catalysts

There are two types of catalyst, **heterogeneous** and **homogeneous**. These are compared in the table below.

	Heterogeneous	Homogeneous
Phase	Different (e.g. gaseous reactants and a solid catalyst)	Same (e.g. reactants and catalyst both in solution or both in gas phase)
Examples	• Hydrogenation of alkenes with a platinum or nickel catalyst • Manufacture of ethanol from steam and ethene using phosphoric acid on a silica support	• Chlorine radicals in the depletion of stratospheric ozone • Enzymes
How does the catalyst work?	• Reactants are adsorbed on to the surface of the catalyst and bonds are weakened • Bonds break • New bonds form • Products desorb from the surface and diffuse away	• Usually one reactant forms an intermediate with the catalyst • The intermediate reacts further to form the products and regenerate the catalyst

Tip At AS, 'phase' is taken to mean the same as 'state'.

Answers to questions

(1) The *f*-orbitals first occur in the fourth period.

(2) (a)

- Metallic elements consist of atoms.
- Non-metallic elements form molecules (noble gases are single atoms) or networks.
- Compounds between metals and non-metals have ionic bonding, and therefore ions are present.
- Compounds between non-metals have covalent bonding and are molecular or networks.
- Acidic or alkaline solutions, formed by adding water to some molecular compounds, contain ions.

(b)

- KCl: ions
- PF_3: molecules
- Ba: atoms
- Poly(propene): molecules
- HBr(aq): ions and molecules (water)

(3)

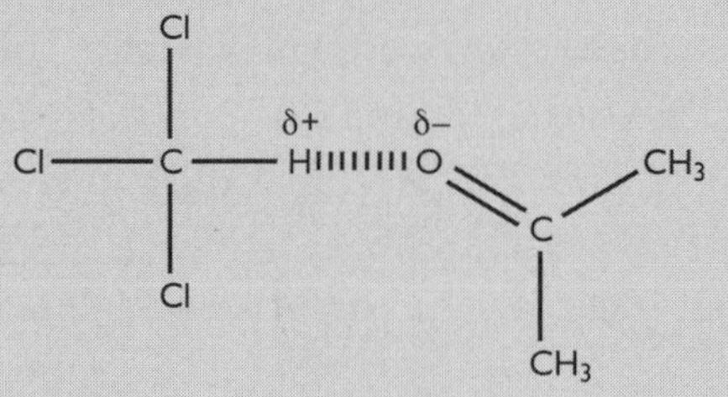

Cl
Cl—C—H ||||||| O=C(CH3)CH3
Cl
δ+ δ−

(4) Sulfur in $Na_2S_2O_3$: +2
Nitrogen in NO_3^-: +5
Phosphorus in P_2O_3: +3
Nitrogen in NH_4^+: –3

(5) Chloride ions have three electron shells (2, 8, 8); sodium ions have two electron shells (2, 8).

(6) (a) The first ionisation enthalpy of magnesium is the energy required to remove one electron from every atom in one mole of isolated gaseous atoms of magnesium; one mole of gaseous magnesium ions with one positive charge are formed.

(b) $Mg^+(g) \rightarrow Mg^{2+}(g) + e^-$

(7) (a) As you go down group 2, the number of protons in the nucleus increases and the shielding effect of the inner shells of electrons keeps the 'effective' nuclear charge approximately the same for each element. However, going down the group, the outer electrons become further away from the nucleus and are less strongly attracted, so it is easier to remove an electron from an atom.

(b) Across a period, electrons are being added to the same shell. At the same time, protons are being added to the nucleus. The increasing positive nuclear charge holds the outer electrons more tightly, so it becomes harder to pull an electron from the outer shell.

(c) The first and second ionisations of magnesium correspond to the removal of two electrons from the outer shell. This is relatively easy for magnesium when it forms compounds. However, the third ionisation involves removing an electron from a filled inner shell. This filled shell is closer to the nucleus and the electrons are held much more tightly, so it requires a very large amount of energy, which is not available for forming compounds containing M^{3+} ions.

(8) (a) The colourless solution turns yellow-brown (the exact shade depends on the concentration of the potassium iodide solution and the amount of chlorine added).

(b) $2I^-(aq) + Cl_2(aq) \rightarrow 2Cl^-(aq) + I_2(aq)$

(9) (a) Organic liquids and solutions do not mix with aqueous solutions, so the two immiscible layers can be separated using a tap-funnel.

(b) With $NaHCO_3$ to remove any acid; with water to remove any $NaHCO_3$.

(c) Any unreacted butan-1-ol.

(10) The C–H bond can easily be broken by radicals present in the atmosphere. Such compounds are broken down in the lower atmosphere and so fail to reach the stratosphere.

(11) In hydrolysis, water breaks down a substance, forming at least two products. In hydration, water adds to a compound, forming a single substance.

(12) (a) In nucleophilic substitution, a hydroxide ion donates a pair of electrons to the electron-deficient carbon bonded to the bromine atom. A covalent bond is formed and at the same time the C–Br bond breaks heterolytically to form a bromide ion and an alcohol.

(b) In electrophilic addition, as the bromine molecule approaches ethene, the high electron density of the C=C bond polarises it, with the nearer bromine atom becoming slightly positively charged. This electrophilic bromine accepts an electron pair from the C=C bond, forming a C–Br covalent bond. At the same time the Br–Br bond breaks heterolytically, forming a bromide ion and a carbocation intermediate. The bromide ion is attracted to the carbocation and a covalent bond is formed.

(13) To break the carbon–halogen bond, the activation enthalpy for the hydrolysis of halogenoalkanes must be exceeded. A C–F bond is much stronger than a C–Cl. Chloroalkanes are therefore more easily hydrolysed.

(14) The ring prevents the single C–C bond between the two Br atoms from rotating.

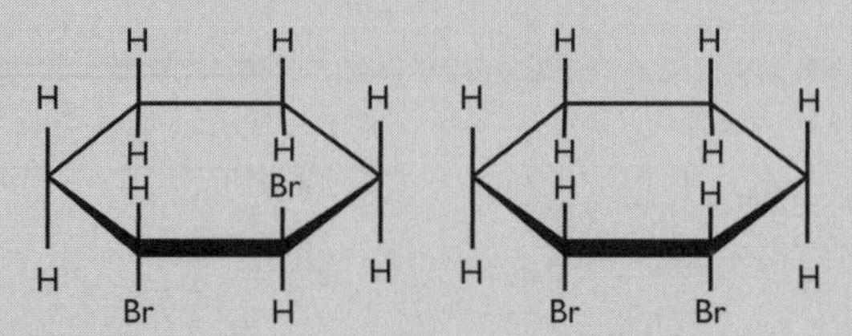

(15) A is bromoethane; B is 2,3-dimethylbut-2-ene.

(16) A C–F bond is stronger than a C–Cl bond. A C–Cl bond requires radiation of a lower frequency than a C–F bond to break it.

(17) A C–F bond is stronger than a C–Cl bond. It needs more energy to stretch it and to cause it to vibrate. A higher infrared frequency is needed, since $\Delta E = h\nu$.

(18) Although methane and CFCs have higher greenhouse factors than CO_2, the concentration of CO_2 is much higher than those of methane and CFCs. Therefore, its overall effect on global warming is more significant.

(19) The strong absorption at about 1700 cm^{-1} indicates a carbonyl group (C=O); the broad absorption peak at about 3100 cm^{-1} indicates a hydroxyl group (OH). Hence, the functional group present in compound **C** must be a carboxyl (COOH).

(20) At the bottom of the ocean, the pressure is greater. This increased pressure causes the equilibrium position for the dissolution of gaseous CO_2 molecules in water to move to the right. More gaseous CO_2 will dissolve.

Questions & Answers

This section contains questions similar in style to those in the Unit F332 examination. The questions begin with a sentence or two to set the context for the various structured parts that follow. This context is called the **stem**.

At AS, some of the contexts of the questions are based on those in the course materials. Others may introduce chemical substances and applications not previously encountered. Either way, these are only vehicles to introduce assessment questions using the chemical knowledge and ideas laid out in the specification. The context can usually be classified under one of the three teaching sections of the unit, and should give you an insight into which aspect of the course is being examined.

Papers are designed to be accessible to candidates who have gained a grade C in double award science at GCSE. Sometimes it is easy to miss the obvious, because only GCSE knowledge is required.

In this guide, the questions in each set are, as far as possible, based on the content of the teaching module involved:

- Set 1: Elements from the sea (ES)
- Set 2: The atmosphere (A)
- Set 3: The polymer revolution (PR)

How to use this section

There are several ways of using these questions and answers. You could read about the appropriate context in your course materials and notes and then tackle the particular question. In extended writing questions, where there is a mark allocated for the quality of the response, this is indicated by the appropriate statement. The answers given are model responses. Together with the examiner's comments, they should enable you to assess your performance. The appropriate learning outcome from the specification is given at the end of each answer.

Examiner's comments

In the answers to questions with more than one marking point, each mark allocation is indicated by a ✓. Where appropriate, the answers are followed by examiner's comments. These are preceded by the icon e. They point out where credit is due and highlight specific problems and common errors.

Elements from the sea

Question 1

The production of bromine

In one process for extracting bromine from sea water, the sea water is first concentrated, by using solar energy to crystallise out sodium chloride. The concentration of bromide ion in the resulting solution is at least 0.028 mol dm^{-3}.

(a) (i) What volume of the concentrated solution is needed to produce at least 1 kg of bromine? (3 marks)

(ii) Draw a structure to show how the ions in crystalline sodium chloride are arranged. In your drawing, show how many chloride ions surround a given sodium ion, the relative sizes of the two ions, and their charges. (3 marks)

(b) In the first stage of the process, the concentrated bromide solution is treated with chlorine.

(i) Describe what is seen in the reaction. (1 mark)

(ii) Write an ionic equation for the reaction of chlorine with bromide ions in sea water. Include state symbols in your equation. (2 marks)

(c) The bromine concentration of the water is very low and is increased by further processing. This results in the formation of a solution of hydrobromic acid, HBr.

A chemist performed a titration experiment to determine the concentration of a solution of hydrobromic acid using 25.0 cm^3 samples of 0.0100 mol dm^{-3} aqueous sodium hydroxide. The chemist found that the samples of the aqueous alkali were exactly neutralised by 22.5 cm^3 of the hydrobromic acid solution.

The equation for the reaction between hydrobromic acid and sodium hydroxide is given below:

$HBr + NaOH \rightarrow NaBr + H_2O$

(i) What did the chemist use to measure out the samples of aqueous sodium hydroxide? (1 mark)

(ii) How did the chemist determine the end-point of the titration? (1 mark)

(iii) Calculate the amount, in moles, of NaOH used in the titration. (2 marks)

(iv) Calculate the amount, in moles, of HBr in the hydrobromic acid solution needed to exactly neutralise 25.0 cm^3 of the aqueous sodium hydroxide. (1 mark)

(v) Calculate the concentration of the hydrobromic acid. (2 marks)

(d) In the final stage of the manufacturing process, the bromine is regenerated using a redox reaction. Suggest a suitable reagent for treating the hydrobromic acid. (1 mark)

Total: 17 marks

Answers to Question 1

(a) (i) Amount of Br in 1 kg of bromine = 1000/79.9 = 12.5 mol ✓.

1 dm^3 of solution contains 0.028 mol of Br
Therefore the volume required = 12.5/0.028 dm^3 ✓ = 446 dm^3.
= 450 dm^3 (to two significant figures) ✓ ES(a)

In calculations, give the final answer to the appropriate number of significant figures.

(ii)

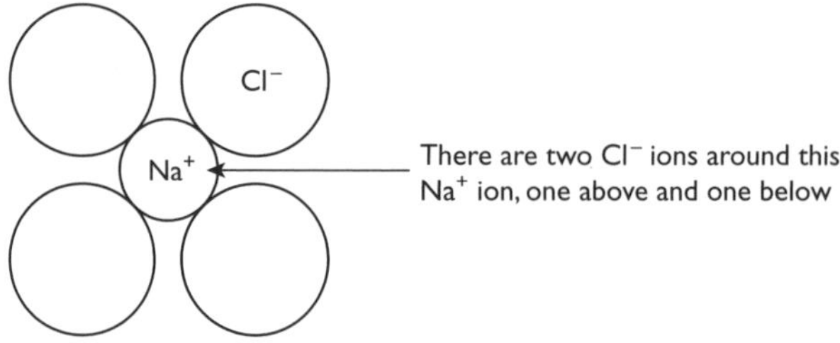

Six Cl^- ions around each Na^+ ion ✓
Relative sizes correct ✓
Charges correct ✓ ES(h)

Sometimes it is easier to draw part of a diagram in two dimensions and add a sentence to make the three-dimensional aspects clear, as in this answer.

(b) (i) An orange solution ✓. ES(m)

The actual colour could be in the range from yellow to red-brown, depending on the concentration. You do not need to describe the colour change, only what colour forms. If a colour *change* is required, it will be asked for specifically.

(ii) $2Br^-(aq) + Cl_2(aq) \rightarrow 2Cl^-(aq) + Br_2(aq)$ ES(o)

There are usually 2 marks awarded for ionic equations. The correct formulae and a balanced equation yield 1 mark ✓; another mark is awarded for the correct state symbols, in this case $Cl_2(aq)$ and $Br_2(aq)$ ✓.

(c) (i) A volumetric/bulb pipette ✓.

(ii) By adding a suitable indicator and looking for the correct colour change ✓.

If you name a specific indicator, make sure you give the correct colour change, e.g. phenolphthalein turns colourless, *not* pink, because acid is being added to alkali.

(iii) Amount of sodium hydroxide added = 0.0100 × 25.00/1000 mol ✓.
= 0.00025 mol ✓.

The first mark is awarded for having moles = concentration × volume.

(iv) Amount of bromide ion added = 0.00025 mol ✓.

(v) Concentration of HBr = 0.00025/(1000/22.5) ✓.
= 0.0111 mol dm^{-3} ✓. ES(b)

(d) Chlorine ✓. ES(o)

For industrial processes, please choose the obvious and cheapest chemical. In this case it is already being used.

Question 2

The manufacture of vinyl chloride

Large amounts of vinyl chloride are needed for the production of the polymer, PVC. The flow diagram summarises the stages involved in the production of vinyl chloride, $CH_2=CHCl$, from ethene and chlorine.

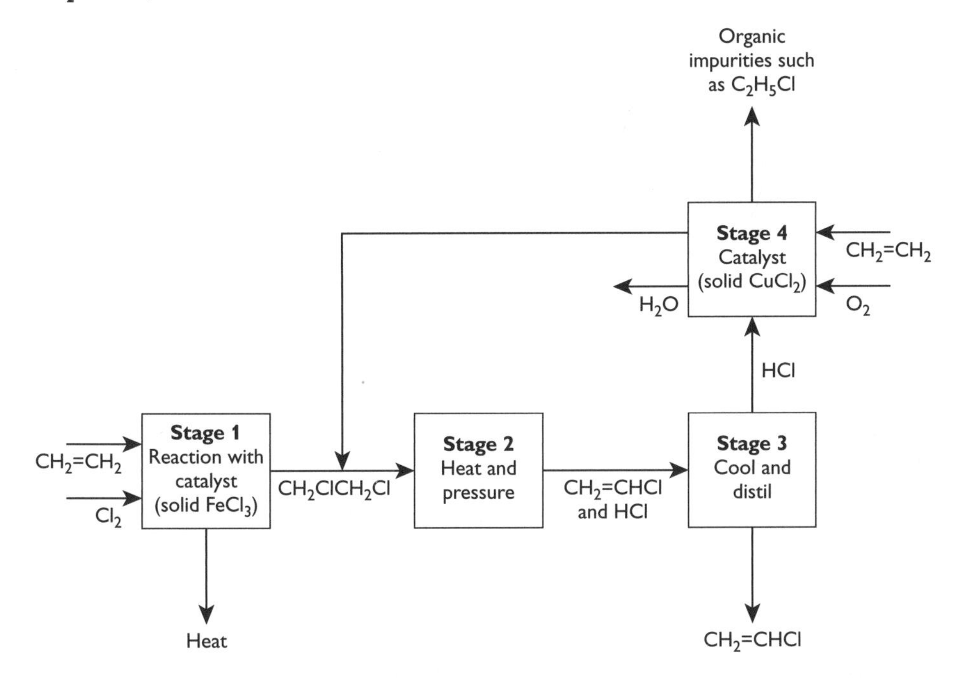

(a) What is the systematic name for vinyl chloride? (1 mark)

(b) (i) In Stage 1, what type of reaction is taking place? (1 mark)

(ii) The chlorine used in the reaction is usually made on site. Other than cost, suggest and explain one reason for this. (1 mark)

(iii) The reaction in Stage 1 is very exothermic. Suggest two ways in which this might benefit the overall process. (2 marks)

(c) Two of the chlorine-containing compounds produced in the process have the boiling points shown in the table below:

Compound	Boiling point (°C)
C_2H_5Cl	12
CH_2ClCH_2Cl	84

In each case, identify the type(s) of intermolecular bonding present between the molecules, and explain the difference in the values of their boiling points.

(3 marks)

(d) What is the purpose of cooling the products from Stage 2? Give a reason for your answer. (2 marks)

(e) (i) Explain how Stage 4 is vital to the overall cost-effectiveness of vinyl chloride manufacture. (2 marks)

(ii) Write a balanced equation for the main reaction occurring in Stage 4. (1 mark)

(iii) Before the water produced in the reaction can be allowed to join a reservoir, it has to be treated. Suggest a reason for this, stating what treatment is necessary. (2 marks)

(f) Both Stage 1 and Stage 4 use catalysts containing *d*-block elements.

(i) Complete the boxes below showing the electron configuration of iron atoms. (2 marks)

1s	2s	2p	3s	3p	3d	4s
↓↑	☐	☐☐☐	☐	☐☐☐	☐☐☐☐☐	☐

(ii) Use your answer to (i) to explain why iron is classified as a *d*-block element. (1 mark)

(iii) Explain, in general terms, how the catalysts involved in Stage 1 and Stage 4 increase the rate of reaction. In your answer you should make clear how your explanation links with the chemical theory. (5 marks)

Total: 23 marks

Answers to Question 2

(a) Chloroethene ✓ ES(t)

(b) (i) Addition ✓ PR(l)

ℯ 'Exothermic' is not an acceptable answer, as it refers to the energy change occurring in a reaction.

(ii) Chlorine is corrosive and toxic, and its transport is a very hazardous operation. ES(q)

ℯ Only one hazardous property of chlorine is required.

(iii) Two points from the following:

- It can be used as a source of energy in Stage 2 ✓.
- It can be used as heat for distillation ✓.
- It can be converted to electricity for chlorine production by electrolysis ✓. ES(n)

ℯ This is not recall knowledge; it is the application of principles found in *Chemical ideas*, Chapter 15.3. You need to consider the information given in the question, including the flow diagram.

(c) Both molecules have instantaneous dipole–induced dipole bonds ✓ and permanent dipole–permanent dipole bonds ✓. These are stronger in CH_2ClCH_2Cl, and more energy is needed to separate the molecules ✓.

> Remember you need to identify the bonds accurately. The use of abbreviations such as 'id–id bonds' will not gain credit. In explaining trends in boiling points and melting points, you must link the strength of bonds between particles to the energy needed to separate those particles. ES(g)

(d) To liquefy the products ✓, so that they can be separated by distillation ✓.

> Remember to give a reason for choosing the answer 'liquefy'. Many students, under examination conditions, forget to do the second part of this type of question.

(e) (i) A large amount of HCl is formed in the process, as a co-product ✓. If the chlorine is not recovered and reused, then the loss of 'chlorine' and the extraction and disposal of the waste HCl will be too great for the overall process to be viable ✓.

> See *Chemical ideas*, Chapters 15.2 and 15.3.

(ii) $2CH_2{=}CH_2 + O_2 + 4HCl \rightarrow 2CH_2ClCH_2Cl + 2H_2O$ ✓

> Credit will be given for equations, if they are balanced using different numbers but the same ratio:
>
> e.g. $CH_2{=}CH_2 + \frac{1}{2}O_2 + 2HCl \rightarrow CH_2ClCH_2Cl + H_2O$

(iii) The water will have some dissolved HCl, or will be acidic ✓, so it will need to be neutralised ✓. ES(a)(n)

(f) (i)

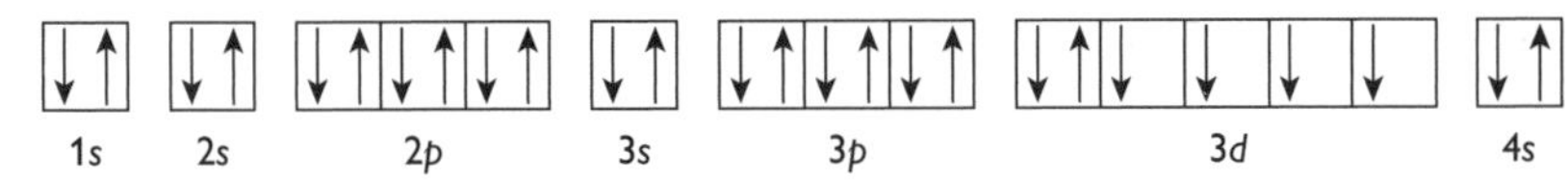

> There is always one easy mark for adding the correct number of electrons. In this case, there are 26 electrons to add. Remember to use your *Data Sheet* to find out the atomic number of the element involved. The second mark is harder to gain and is for the correct arrangement.

(ii) The d sub-shells in iron (Fe) are the final ones being filled ✓. ES(c)(d)

(iii) The reactant molecules are adsorbed on the surface of the catalyst ✓.

The bonds of the reactants are weakened as they interact with the surface ✓. New bonds are formed ✓. Product molecules are desorbed from the surface ✓. *The activation enthalpy/energy for the reaction is lower for the catalytic route* ✓.

> The first four marking points can be phrased in different ways, but as long as the meaning is correct, the mark will be awarded. The final mark is the QWC mark and links how the catalyst works to achieving a reaction pathway with a lower activation enthalpy/energy. A(e)

Question 3

Halogen compounds as anaesthetics

Chloroethane is used as a mild local anaesthetic for treating minor skin wounds. Halothane was developed in the 1950s as a general anaesthetic and was used throughout the world until chemists produced less toxic replacements in the 1990s.

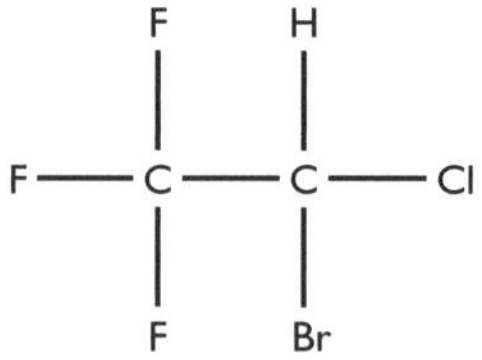

```
      F     H
      |     |
  F — C  —  C — Cl
      |     |
      F     Br
```

Halothane

(a) Chloroethane can be hydrolysed by heating with aqueous sodium hydroxide.

- **(i) Explain the term 'hydrolysed'.** (1 mark)
- **(ii) Draw the full structural formula for the organic product produced when chloroethane is hydrolysed.** (1 mark)
- **(iii) Using curly arrows, describe the mechanism of the hydrolysis of chloroethane by the hydroxide ion.** (3 marks)

(b) A student reacted a sample of chloroethane with a warm mixture of ethanol and aqueous silver(I) nitrate(V) solution in a test-tube until hydrolysis occurred.

- **(i) Describe what the student saw when hydrolysis took place.** (2 marks)
- **(ii) Write an ionic equation for this reaction. Include state symbols in your equation.** (3 marks)

(c) (i) Give the systematic name of halothane. (2 marks)

- **(ii) Some data for the halogens present in halothane are given in the table below.**

Halogen (X)	C–X bond enthalpy(kJ mol^{-1})	Electronegativity
Fluorine	452	4.0
Chlorine	346	3.0
Bromine	290	2.8

Identify the halogen in halothane that hydrolyses first when the hydroxide ion is warmed with halothane. Explain how both sets of data might be used to predict which atom would hydrolyse first. State which set of data is the more important. In your answer you should make clear how your explanation links with the chemical theory. (5 marks)

(d) One of the replacement compounds for halothane is isoflurane.

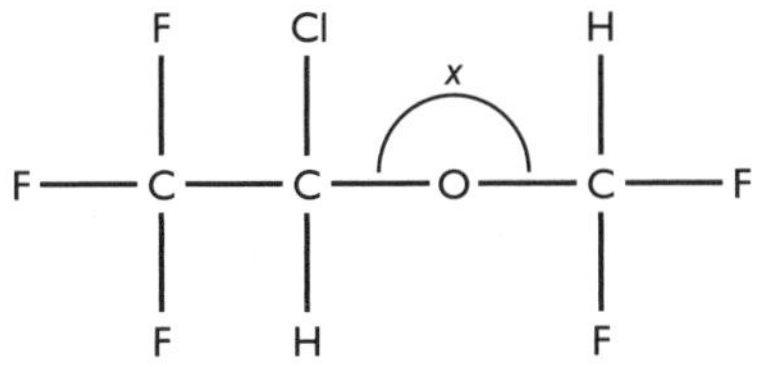

Isoflurane

(i) Name the functional group present in isoflurane but not in halothane. (1 mark)

(ii) Use the electron pair repulsion principle to predict a value for the bond angle, labelled *x*, in the structure of isoflurane. (3 marks)

Total: 21 marks

Answers to Question 3

(a) (i) It is broken down into at least two substances by reaction with water ✓.

(ii) ✓

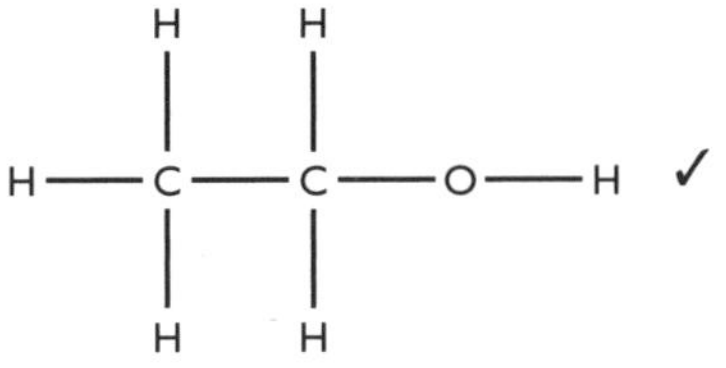

ES(x) ES(w) ES(aa)

Remember to draw the OH group showing its bond. You are not asked about bond angles, so don't try to draw the structure in three dimensions. Keep it simple.

(iii)

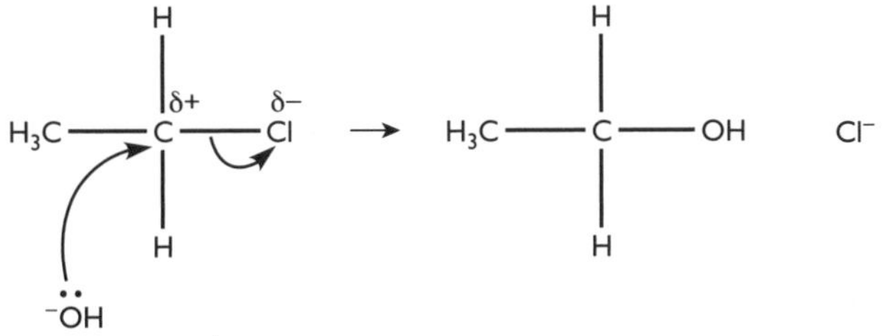

H_3C—$C^{\delta+}$—$Cl^{\delta-}$ → H_3C—C—OH Cl^-

^-OH

The answer should give the correct formulae for the molecules and ions ✓.
The curly arrows should begin and end at the correct points ✓.
Partial charges should be shown correctly ✓. ES(y)

(b) (i) A white ✓ precipitate ✓.

(ii) $Ag^+(aq) + Cl^-(aq) \rightarrow AgCl(s)$

In this equation, there is 1 mark for the correct formula for silver ion, 1 mark for the rest of the equation being correct and 1 mark for the correct state symbols. A common error is to write Ag^{2+}. If you do this but the rest is balanced correctly for a 2+ charge, you will gain 2 out of the 3 marks.

(c) (i) 2-Bromo-2-chloro-1,1,1-trifluoroethane
The correct chain and substituents should be given ✓.
The correct numbers should be given ✓.

This question is hard, but it does illustrate the rules for naming halogenoalkanes.

(ii) The bromine atom is first to be replaced by an OH group ✓.

The electronegativity values indicate that a C–F bond is the most polar ✓, and if this factor is more important, then attack by the nucleophile will occur more easily at the carbon of the C–F bond ✓. If the bond enthalpy is more important, then the weakest bond, C–Br, will break first ✓. Experimental results prove that bond enthalpy is the more important factor ✓.

The final mark is the QWC mark and this links the experimental result (hydrolysis in the presence of silver(I) nitrate(V) solution) to bond enthalpies. ES(z)

(d) (i) Ether ✓ ES(s)

(ii) There are two bonding pairs of electrons ✓ and two lone pairs around the O atom ✓. Repulsion of the electron pairs leads to a tetrahedral shape for the orientation of electron pairs, so the bond angle, *x*, will be approximately 109° ✓.
ES(e)

The atmosphere

Question 1

CFCs

Although CFCs have been banned from being used as solvents because of their effect on the ozone layer, other chlorine compounds such as dichloromethane are still being widely used.

(a) (i) Draw a three-dimensional structure for dichloromethane. (1 mark)

(ii) Use the electronegativity data given below to deduce whether or not dichloromethane is a polar molecule. Give your reasoning.
Electronegativity: C, 2.6; H, 2.2; Cl, 3.2 (2 marks)

(b) Dichloromethane can be made from methane and chlorine in the presence of UV radiation. Chloromethane is formed first and further reaction produces dichloromethane.

(i) The mechanism of the reaction is a radical chain reaction.
Write equations for the initiation step and for the two propagation steps for the formation of dichloromethane from chloromethane. (3 marks)

(ii) A small amount of 2,3-dichlorobutane is formed during the reaction. Suggest a termination step to account for this. (1 mark)

(c) Explain how CFCs cause the breakdown of the ozone layer and why dichloromethane is less of a problem. Include equations, showing how a large amount of ozone is destroyed by a small number of CFC molecules. (7 marks)

(d) The formulae for three replacement compounds for CFCs are shown below.

$CHClF_2$	CH_3CF_3	C_4H_{10}
A	**B**	**C**

(i) All three compounds are considered unsatisfactory in the long term, for the same reason. What is this disadvantage? (1 mark)

(ii) Compounds such as A and B decompose easily in the troposphere by reaction with hydroxyl radicals. Write an equation for the reaction of compound A with a hydroxyl radical. (1 mark)

(iii) Compound C is often used in aerosol sprays. Why might such sprays be dangerous? (1 mark)

Total: 17 marks

Answers to Question 1

(a) (i)

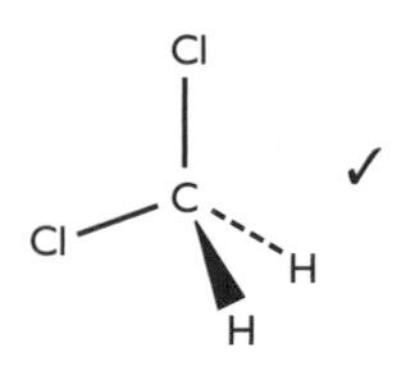

(ii) The C–Cl bond is polar because it has an appreciable electronegativity difference of 3.2 – 2.6 = 0.6 ✓. Since the C–Cl dipoles do not cancel each other out, due to the shape of the molecule, the molecule will be polar ✓.

There is 1 mark for using the data and 1 mark for deducing why the molecule is polar. Often this type of question can score more marks if the partial charges are requested or the molecule's shape needs to be considered because there is more than one polar bond. Remember that C–H bonds are considered to be essentially non-polar, even though the electronegativity difference is 0.4. ES(aa) ES(f)

(b) (i) Initiation:

$Cl_2 \rightarrow 2Cl$ ✓

Propagation:

$C_2H_5Cl + Cl \rightarrow C_2H_4Cl + HCl$ ✓

$C_2H_4Cl + Cl_2 \rightarrow C_2H_4Cl_2 + Cl$ ✓

(ii) $2C_2H_4Cl \rightarrow CH_3CHClCHClCH_3$ ✓

This is a difficult question and requires a good knowledge of chain reactions to answer it. In termination steps, two radicals combine together, hence two different radicals will give rise to three possible termination reactions. A(n) A(o)

(c) In the upper atmosphere (or stratosphere) ✓ UV (or high-energy/high-frequency) radiation ✓ causes CFC molecules to break down to form chlorine radicals ✓. Chlorine radicals catalyse the decomposition of ozone ✓/one chlorine radical can break down many ozone molecules ✓

$Cl + O_3 \rightarrow ClO + O_2$ ✓

$ClO + O \rightarrow Cl + O_2$ ✓

Dichloromethane molecules are broken down in the lower atmosphere (or troposphere) by radicals abstracting hydrogen atoms ✓. A(k)A(q)

(d) (i) They are all greenhouse gases ✓

(ii) $CHClF_2 + OH \rightarrow CClF_2 + H_2O$ ✓

(iii) Butane is flammable and is particularly dangerous when used in hairsprays ✓. A(p)

Question 2

The greenhouse effect and global warming

A Russian scientist has recently proposed an alternative theory to explain the increase in the temperature of the Earth's atmosphere that began between 1906 and 1909. There had previously been a slight decrease in temperature. He suggested that a massive meteorite impact in 1908, which rocked a remote part of Siberia and caused an enormous explosion, resulted in a disturbance in the high-altitude clouds of ice crystals. This produced a large increase in the amount of water vapour in the troposphere (lower atmosphere), thus triggering an increase in global warming.

(a) Explain, using chemical ideas, how an increase in the concentration of water vapour in the troposphere could lead to an increase in the temperature of the troposphere. In your answer, you should use appropriate technical terms, spelled correctly. (6 marks)

(b) Some scientists also argue that the increase in carbon dioxide emissions is not significant, since large amounts of this carbon dioxide dissolve in the oceans.

(i) Describe one cause of the increasing concentration of carbon dioxide in the atmosphere. (1 mark)

(ii) The carbon dioxide in the atmosphere is in equilibrium with carbon dioxide in the surface water of the oceans. Some of this carbon dioxide reacts with the water. Use equations 1 and 2 and Le Chatelier's principle to explain how the oceans reduce the concentration of carbon dioxide in the atmosphere.

equation 1 $CO_2(g) \rightleftharpoons CO_2(aq)$

equation 2 $CO_2(aq) + H_2O(l) \rightleftharpoons HCO_3^-(aq) + H^+(aq)$ (5 marks)

(iii) The presence of H+(aq) ions causes an aqueous solution to be acidic. What other information do you need to know about the reaction described by equation 2 to predict the effect of temperature on the acidity of the water? Explain your reasoning. (3 marks)

(iv) Give the name of the HCO_3^- ion. (1 mark)

(c) Suggest two ways that carbon dioxide emissions could be controlled. Give one advantage and one disadvantage of each approach. (4 marks)

Total: 20 marks

Answers to Question 2

(a) Some UV and visible radiation from the Sun is absorbed (*spelled correctly*) by the Earth ✓. The Earth emits ✓ infrared radiation ✓, some of which is absorbed ✓ by water molecules in the atmosphere. This radiation causes the water molecules to vibrate ✓ faster ✓. This extra energy is transferred to other molecules by collisions ✓. Thus the average kinetic energy of all molecules in the atmosphere increases and the temperature rises ✓.

In the answer, there are eight marking points identified, yet there are only 6 marks allocated. Often this is the case in the extended-writing answers. It is often possible to convey a full understanding without making every feasible point. For instance, some students may start with the Earth, rather than the Sun. The 'faster' mark can also be gained by writing 'causes an increase in (kinetic) energy'. The QWC marking point is awarded for using and spelling 'absorbed' correctly. Many students gain few if any marks on this type of question; there is a tendency to rely on general knowledge for this type of topic. Unfortunately, this usually comes from popular texts and papers and involves such phrases as 'reflects like a mirror', 'bounces off' or 'acts as a blanket'. Use chemical terminology and ideas logically and get the processes involved clear in your own mind. It is not possible to learn such topics parrot-fashion. Six marks out of 90 can be significant. A(s) A(t) A(v)

(b) (i) Increased use of fossil fuels ✓

Deforestation is also valid. There is no need to give lots of detail about the type and use of a particular fuel, or details about photosynthesis.

(ii) An increase in the concentration ✓ of CO_2 in the atmosphere will cause ✓ the equilibrium position of equation 1 to move to the right ✓, increasing the concentration of CO_2 in the oceans. This will now cause ✓ the equilibrium position in equation 2 to also move to the right ✓, thus lowering the proportion of CO_2 in the air ✓. A(h)

There are six marking points and 5 marks to be gained. The first point about the concentration of CO_2 is important in understanding chemical equilibrium. It is therefore allocated 1 mark. There are 4 marks for any four of the remaining five marking points.

Remember to use the numbers for the equations, as it saves a lot of description. Do *not* restate the equations in words as the basis for your answer. Many students do this, and fail to gain any credit. This is always one of the worst-tackled concepts in examinations at this level. Be careful.

(iii) Whether the reaction is exothermic or endothermic ✓. If the reaction is exothermic, then an increase in temperature will move the equilibrium position to the left ✓. The hydrogen ion concentration will therefore decrease, thus lowering the acidity ✓.

The first mark can be also gained by giving the enthalpy change for the reaction. For the second mark, Le Chatelier's principle is applied. You don't need to waste space and time quoting the principle — you should just show that you know how to use it. Finally, you need to use the information given in the question about acidity. A(h)

(iv) Hydrogencarbonate ✓

(c) The marks are for two of the following three alternatives:

Method	**Disadvantage ✓**	**Advantage ✓**
Burn fewer fossil fuels	There could be an energy shortfall Or: It is expensive to develop alternative fuels	Fossil fuels will last for a longer period of time
Increase photosynthesis	Land shortages for growing population Or: It is expensive and difficult to develop synthetic methods of photosynthesis	Little maintenance is required. Or: Synthetic photosynthesis could be used where vegetation is difficult to grow
Burying carbon dioxide	Expensive	At the bottom of the oceans it will stay there as a liquid for ever

e The marks are for describing the advantages and disadvantages for each approach. This is a topic which is continuously changing, and you will be given credit for any reasonable answer that you researched during the course. A(x)

Question 3

The catalytic decomposition of N_2O

Dinitrogen oxide, N_2O, is an anaesthetic analgesic drug, sometimes called 'laughing gas'. It is a very potent greenhouse gas and is 300 times more effective than carbon dioxide. Dinitrogen oxide is produced naturally, and so it is present in significant amounts in the atmosphere, where it is comparatively inert. It only decomposes to nitrogen and oxygen at temperatures in excess of 1 000 K. However, the decomposition is catalysed by traces of chlorine gas. It is an example of a homogeneous catalysed reaction.

(a) (i) What does the word *homogeneous* say about the catalyst in this reaction? (1 mark)

(ii) Draw a labelled enthalpy-profile diagram to show how a catalyst speeds up a chemical reaction. (5 marks)

(b) The catalyst is thought to be chlorine radicals formed by the homolytic photodissociation of chlorine gas. The proposed mechanism is:

Step 1: $N_2O(g) + Cl(g) \rightarrow N_2(g) + ClO(g)$

Step 2: $2ClO(g) \rightarrow Cl_2(g) + O_2(g)$

(i) Explain the terms *radical* and *homolytic photodissociation*. (3 marks)

(ii) Give two examples of radicals, other than Cl, from the equations for Steps 1 and 2. (1 mark)

(iii) Combine the equations for Steps 1 and 2 to give the overall equation for the decomposition of dinitrogen oxide. (1 mark)

(c) Chlorine atoms are formed when chlorine molecules absorb ultraviolet radiation of the appropriate frequency.

(i) The bond enthalpy of a Cl–Cl bond is +243 kJ mol^{-1}. Calculate the energy in joules needed to break a Cl–Cl bond in a single molecule. The Avogadro constant is 6.02×10^{23} mol^{-1}. (2 marks)

(ii) Use $\Delta E = h\nu$ to calculate the minimum frequency needed to break a Cl–Cl bond.
$h = 6.63 \times 10^{-34}$ J Hz^{-1} (2 marks)

(d) Dinitrogen dioxide requires a temperature of 1000 K to decompose without the presence of a catalyst. Explain, using collision theory, why the reaction requires a temperature of 1000 K. In your answer, you should make clear how your explanation links with the chemical theory. (6 marks)

Total: 21 marks

Answers to Question 3

(a) (i) It is in the same phase as the reactants, in this case, the gas phase ✓.

'State' instead of 'phase' is acceptable.

(ii)

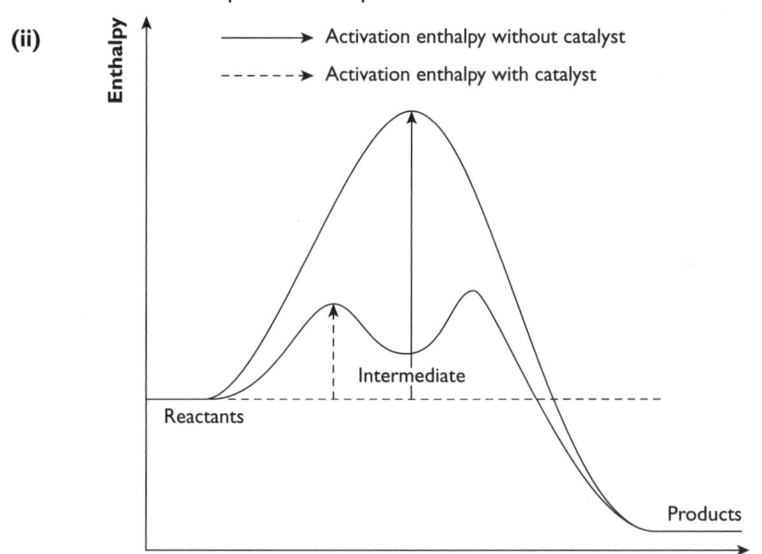

This question is often asked by giving the student a diagram to complete. The way that the question is asked here is harder, because you have to remember what an enthalpy diagram is and then label your own drawing. I would allocate the marks as follows: 1 mark for labelling the activation enthalpies correctly. At AS, double-headed arrows are acceptable, but not at A2. One mark is also awarded for showing that the activation enthalpy with the catalyst is lower than for the uncatalysed reaction.

One mark is then given for two 'humps' on the profile for reaction with a catalyst. Finally, 2 marks from the three marking points are awarded for: labels for the axes; labelling the reactants and products correctly; and labelling the 'trough' as intermediate.

Remember that you have not been asked to describe the diagram in words or to explain how a catalyst works. Without a diagram there will be no credit given.

A(c) A(e) A(f)

(b) (i) A *radical* is a chemical species (atoms, ions or molecules) that has at least one unpaired ✓ electron. *Homolytic photodissociation* is the breaking of a covalent bond into two radicals ✓ by the absorption of light ✓.

'Spare' or 'lone electron' are acceptable, but *not* 'free', 'unshared' or 'single electron'.

(ii) ClO and O_2 ✓

Don't forget that oxygen molecules are biradicals.

(iii) $2N_2O \rightarrow 2N_2 + O_2$ ✓ A(m) A(n)

(c) (i) energy of Cl–Cl bond = $+243 \times 10^3/6.02 \times 10^{23}$ J per bond ✓
$= 4.04 \times 10^{-19}$ J ✓

(ii) $\nu = 4.04 \times 10^{-19}/6.63 \times 10^{-34}$ Hz ✓
$\nu = 6.09 \times 10^{14}$ Hz ✓ A(u)

Remember that the space for the answers on the exam paper will show the units for energy (J) in part (i) and frequency (Hz) in part (ii). However, many students forget to convert kilojoules into joules, which would lose one of the marks. Try to remember that visible and ultraviolet frequencies are around 10^{14} to 10^{15} Hz.

(d) For particles to react, they must collide ✓. *For product molecules to form, the collision must have sufficient energy to do so* ✓. This energy is called the activation enthalpy for the reaction ✓. At higher temperatures the average kinetic energy of the particles is greater ✓. More collisions will therefore have the necessary activation enthalpy ✓, and more product molecules will form, thus increasing the rate of reaction ✓.

Remember that it is the particles undergoing the collision which must have sufficient energy — not just one of them. Students often miss the final marking point, relying on examiners to infer that this is what the students meant to say.

The marking point in italics is the QWC mark, and this links the reaction rate to the activation enthalpy. A(d)

The polymer revolution

Question 1

PVC

PVC was discovered by accident in the nineteenth century. When glass flasks containing chloroethene were left in sunlight, a white solid formed on the glass surface. Even though the polymer was a thermoplastic and not a thermoset, the solid was very brittle and difficult to work. In the 1920s, chemists were able to make it more flexible by adding compounds that acted as plasticisers and it quickly became widely used in a variety of products.

(a) (i) Draw a structural formula for part of a PVC chain consisting of three monomer units. (1 mark)

(ii) Describe and explain the difference between a thermoplastic polymer and a thermoset. (3 marks)

(iii) Suggest a use for each of the two types of PVC, explaining your choice of polymer. (1 mark)

(b) The structure below shows part of a copolymer chain formed by polymerising chloroethene with compound P.

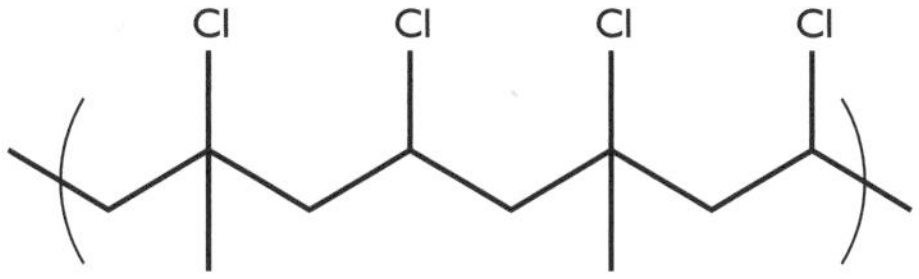

Cl Cl Cl Cl

Draw the full structural formula of compound P. (1 mark)

(c) Environmental groups have called for the global phasing out of PVC, because it is claimed that dioxins are produced during its manufacture and by burning waste PVC. Dioxins are known to increase the risk of cancer.
The structure of 1,4-dioxin is shown below.

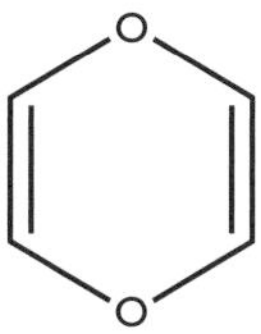

O
O

(i) Describe a chemical test to show that 1,4-dioxin is unsaturated. State what you would see. (2 marks)

(ii) Choose two words from the following list to describe the mechanism of the reaction between an alkene and hydrogen bromide:

radical, electrophilic, nucleophilic, redox, addition, substitution, elimination (2 marks)

(iii) Complete the diagram below, using curly arrows, to show the mechanism of the reaction between 1,4-dioxin and bromine. Name the type of intermediate formed.

O O C=C H H Br–Br → Intermediate → Compound A

(4 marks)

(iv) Compound A forms *E*/*Z* isomers. Suggest and explain how it is able to do this. (2 marks)

(v) Name the functional group, other than an alkene, that is present in 1,4-dioxin. (1 mark)

(d) Give the reagent and conditions for converting 1,4-dioxin into compound B.

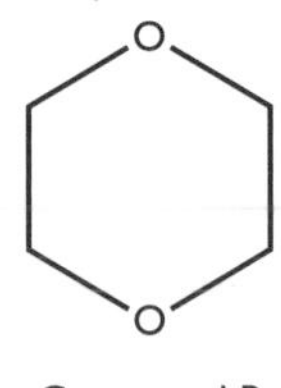

Compound B

(2 marks)

Total: 19 marks

Answers to Question 1

(a) (i)

Cl Cl Cl

✓

The brackets are not essential, but you have to give the bonds linking the three monomer units to further units. PR(i)

(ii) Thermosetting polymers do not soften on heating, whereas thermoplastic polymers do ✓. In a thermoplastic polymer, the intermolecular bonds between chains are weak ✓. In a thermosetting polymer, the intermolecular bonds between chains are strong covalent bonds ✓.

e The word 'covalent' in the final marking point is essential. PR(o)

(iii) Unplasticised PVC is used in the construction industry because of its rigidity, e.g. for wall cladding, fences and window frames ✓.
Plasticised PVC is used for packaging or as a fabric, because of its flexibility, e.g. for plastic bottles or plastic raincoats ✓.

e Uses given without a reason will not score any marks. In the specification, PR(n), you are expected to be able to suggest uses for polymers, based on their properties.

(b)

H–C(H)=C(Cl)–CH₃ drawn with all bonds shown: H, C, H, C, Cl, H—C—H, H ✓

e Remember to draw the CH_3 group showing the individual C–H bonds. Check the numbers of bonds around each type of atom, e.g. every carbon has four bonds, and where there is oxygen every oxygen has two bonds while halogens have a single bond. ES(aa)

(c) (i) Shake the 1,4-dioxin with a few drops of bromine ✓.
The solution will change from orange to colourless ✓.

e Do not use the phrase the 'solution turns clear'. Remember that all solutions are clear, but not all are colourless. PR(j)

(ii) Electrophilic ✓ addition ✓ PR(l)

(iii)

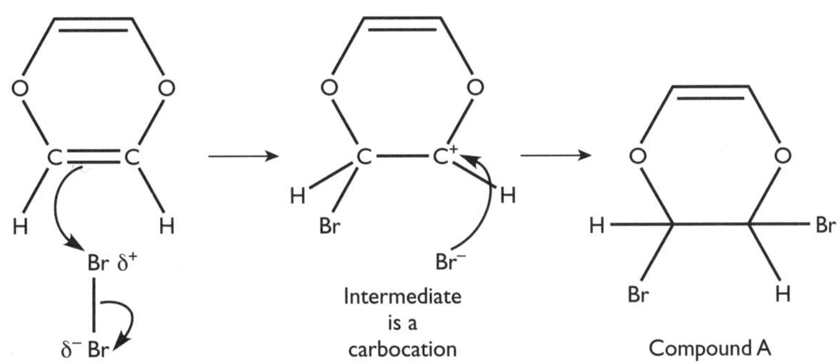

Curly arrows should be correct for the reactants ✓.
Partial charges should be shown on the Br_2 ✓.

The structure of the intermediate should be correct ✓.
The bromide ion should be shown attacking ✓ the carbocation ✓.

(iv) The ring structure prevents free rotation around the C–C bond ✓, so that the two bromine atoms can be either on the same side, or on opposite sides. ✓ PR(m)

The use of the words 'suggest and explain' in the question indicates that this is an application of a known chemical idea to an unusual example.

(v) Ether ✓ ES(aa)

(d) Hydrogen ✓ with *either* a Ni catalyst, heat and pressure, *or* a Pt catalyst at room temperature and pressure ✓. PR(j)

Question 2

Polymers that dissolve in water

The diagram below shows the structure of part of polymer chain. The polymer is commonly known as polyvinyl alcohol, PVOH, and has water-soluble properties.

(a) (i) **Draw the structure of the monomer used to make PVOH.** (1 mark)

(ii) **Give the systematic name for PVOH.** (1 mark)

(b) **The monomer of PVOH is unstable and rearranges to form compound A.** (1 mark)

Compound A

(i) **Name the functional group present in compound A.** (1 mark)

(ii) **A student tried to synthesise the monomer of PVOH. The IR spectrum of the product is shown below. Show how this spectrum confirms that the student has created compound A rather than the monomer.**

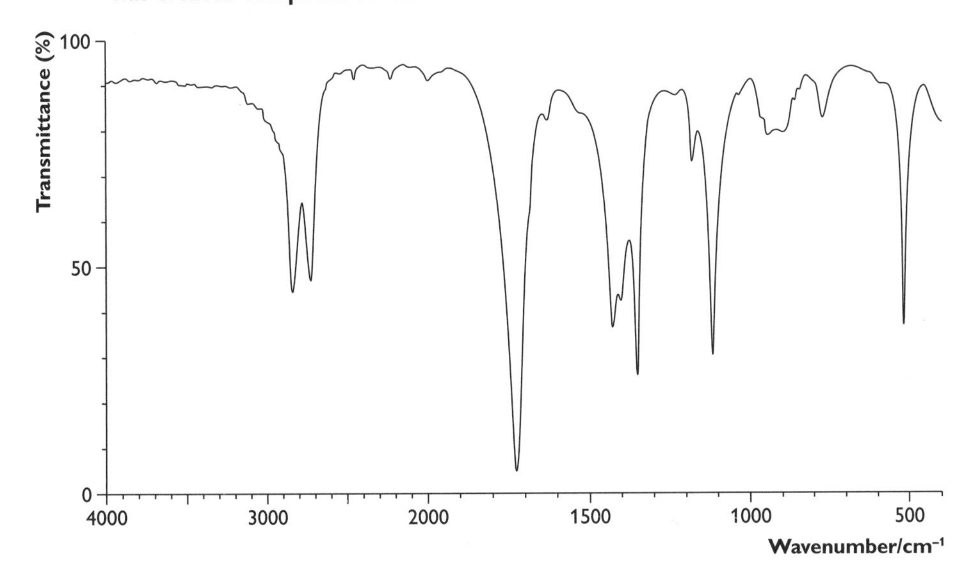

(3 marks)

(c) PVOH is made by first polymerising compound B.

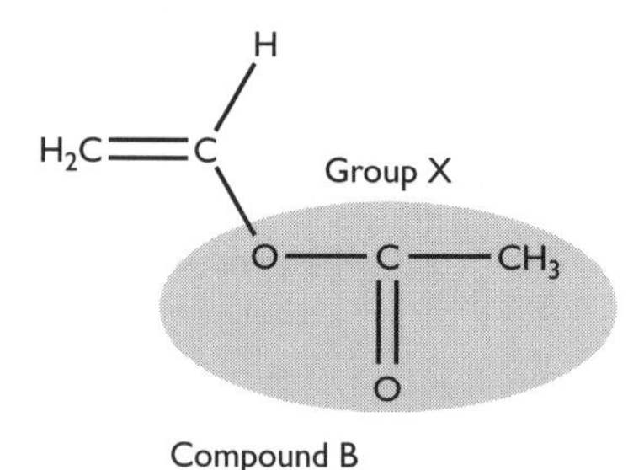

Compound B

The resulting polymer is insoluble in water. Its functional groups, labelled X, can be replaced with OH groups. The polymer's solubility in water depends on the percentage of OH groups. Polymers with up to 90% OH groups are soluble in cold water, but pure PVOH is insoluble.

(i) What name is given to the type of reaction in which compound B is polymerised? (1 mark)

(ii) Draw the structural formulae of two water molecules, showing how they can form a hydrogen bond. Indicate clearly any partial charges. (3 marks)

(iii) Explain, in terms of intermolecular forces, why the water solubility of the polymer varies with the percentages of X and OH groups. Comment on whether polymers with mainly X groups will be soluble or not. In this question, one of the marks is for QWC. (6 marks)

(d) Suggest one use for PVOH film. (1 mark)

Total: 18 marks

Answers to Question 2

(a) (i)

H
H_2C=C ✓
O—H

PR(d) PR(i)

You have not been asked for the 'full' structural formula, so –OH is acceptable here.

(ii) Poly(ethenol) ✓ PR(c)

(b) (i) Aldehyde ✓ PR(e)

The answer 'carbonyl' is not sufficient, and would not score.

(ii) The 'strong' peak at about 1725 cm^{-1} ✓ indicates the presence of a C=O bond ✓ in an aldehyde ✓.

Alternatively, the absence of an OH peak at around 3600 cm^{-1} and a 'medium' peak at 1620–1680 cm^{-1} owing to a C=C double bond would gain the 3 marks. PR(p)

(c) (i) Addition ✓ PR(i)

Many students make the mistake of writing 'additional'.

(ii)

$$\begin{array}{c} \delta+ \quad \delta- \\ O—H\cdots\cdots O—H \\ / \qquad\quad / \\ H \qquad\quad H \end{array}$$

There is 1 mark for the correct shape of the water molecule. Beware: you would be surprised at the number of students in examinations who write it with two oxygen atoms and one hydrogen atom. There is one mark for the hydrogen bond being correct. Try to remember to line up the OH group, interaction and lone pair (it is not necessary to show the electrons) on the oxygen atom. The hydrogen bond is directional. The final mark is for correct use of partial charges. PR(a)

(iii) There is extensive hydrogen bonding between chains in pure poly(ethenol) ✓ making the intermolecular bonds very strong ✓. Too much energy is needed to separate the chains ✓. Replacing a few OH groups by X groups disrupts the hydrogen bonding ✓. As the number of X groups increases, the strength of the intermolecular bonding decreases ✓. The polymer dissolves because the OH groups in the chains can now hydrogen-bond to water ✓. If there are lots of X groups, then hydrogen bonding will be much less extensive and the polymer will be insoluble ✓.

There are seven marking points, but only 6 marks available. The first two points and the final point must be identified, for 3 marks. Then any three of the remaining four marking points score 1 mark each. The key intermolecular force in deciding solubility is hydrogen bonding between polymer chains and water molecules. However, since hydrogen bonds have similar energies, it is important to focus on the *extent* of this bonding in deciding the strength of the overall intermolecular forces. There is no need to mention other types of intermolecular bond, since these are much weaker than hydrogen bonding. The final marking point must be addressed in response to the question asked in the stem. PR(c)

(d) Bags for washing hospital laundry, film packaging for detergents ✓. PR(n)

Other acceptable answers include coatings for capsules containing medicines for slow release, and coatings for seeds to protect them from disease and decay.

Question 3

Alcohols

The fermentation of apples to produce cider produces a mixture of alcohols. The main component is ethanol. The table below gives the formula, M_r and the boiling point of some of the other alcohols formed.

Compound	Formula	M_r	Boiling point (°C)
A	$CH_3CH_2CH_2OH$	60	97
B	$CH_3CH(OH)CH_3$	60	82
C	$CH_3CH_2CH_2CH_2OH$	74	118
D	$(CH_3)_3COH$	74	82
E	$CH_3CH(OH)CH_2CH_3$	74	94
F	$CH_3CH_2CH_2CH_2CH_2OH$	88	138

(a) (i) State which compounds are secondary alcohols. Give your reasoning. (2 marks)

(ii) Some alcohols can be oxidised by heating under reflux with a suitable oxidising agent. Explain the meaning of the term 'heating under reflux'. (2 marks)

(iii) Give the reagent used for oxidising alcohols. (1 mark)

(iv) Give the structural formulae of the organic products formed by oxidising, in the manner described in part (ii), for the following compounds:

- **Compound A**
- **Compound B** (2 marks)

(v) Which alcohol cannot be oxidised in the manner described in part (ii)? Give a reason for your answer. (1 mark)

(b) A student tried to explain the trends in the boiling points of compounds A–F, remembering from studies of straight-chain alkanes that their boiling points increased with M_r. However, this did not seem to be the case here. Explain how the student should have applied ideas of intermolecular bonding to explain the trends in boiling points of compounds A–F. (5 marks)

(c) Compound E, when refluxed with concentrated sulfuric acid, produced a mixture of two structural isomers, G and H, with the molecular formula C_4H_8.

(i) The IR spectrum of one of the isomers formed from compound E is shown below. Show how you would use the spectrum to identify the functional group in isomers G and H.

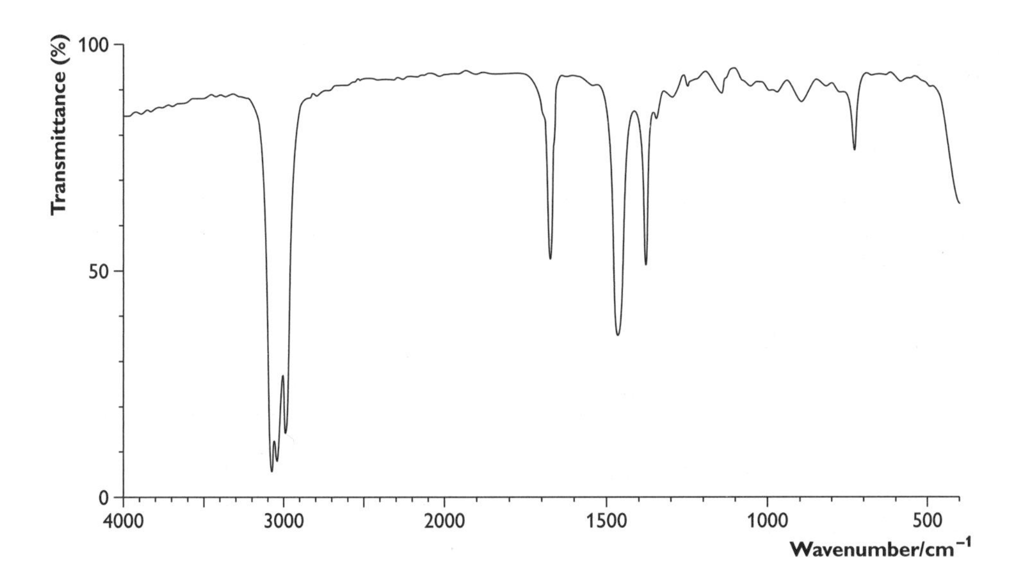

(2 marks)

(ii) Isomer H shows *E/Z* isomerism. Draw the structural formulae of compound G and the two stereoisomers of H. Label each of the two stereoisomers as *cis* or *trans*. (3 marks)

(iii) Give the systematic name of compound G. (1 mark)

(iv) Compound H exists as two stereoisomers. How could their IR spectra show that they are different compounds? (1 mark)

(v) What type of reaction occurs when compound E reacts with concentrated sulfuric acid? Explain your answer. (2 marks)

Total: 22 marks

Answers to Question 3

(a) (i) B and E ✓, because the OH group is attached to a carbon bonded to only one hydrogen ✓.

Alternative answers here are: the OH group is in the middle of a chain, or the OH group is attached to a carbon bonded to two alkyl groups. PR(f)

(ii) Boiling ✓ the mixture but not allowing the vapours to escape, by using a condenser ✓. PR(g)

(iii) Acidified ✓ potassium dichromate solution ✓ PR(k)

The question does not say 'name', therefore formulae are equally acceptable. The formulae may be given as ions, $H^+/Cr_2O_7^{2-}$. The acid may be specified by name, e.g. sulfuric acid.

(iv)

A: H_3CH_2C—C(=O)—OH

B: H_3C—C(=O)—CH_3

There is 1 mark for A and B being correct. Again, you are not asked for full structural formulae, so –OH, or even COOH, is fine. PR(k)

(v) Compound D: it is a tertiary alcohol ✓. PR(f)

(b) The stronger the intermolecular bonding, the higher the boiling point, because more energy is needed to separate the molecules ✓. The straight-chain alcohols, A, C and F, ✓ behave in the same way as the straight-chain alkanes. The boiling points increase because the strengths of the instantaneous dipole–induced dipole bonds increase ✓.

The branching of the chains causes the molecules to have less area in contact with each other (or to be more compact) ✓, thus lowering the strength of the intermolecular bonds ✓.

You need to realise that the hydrogen bonding in each case will be very similar, and that it is the weaker types of intermolecular bonding that are making the difference here. In this type of question, many students forget to link boiling (or melting) point to the energy needed to separate the particles. PR(b)

(c) (i) The medium peak at around $1670\,cm^{-1}$ ✓ indicates the presence of an alkene ✓.

The peaks at around $3000\,cm^{-1}$ may also be used. The peaks just below $3000\,cm^{-1}$ indicate the presence of alkane C–H bonds, and those just above $3000\,cm^{-1}$ indicate alkene or arene C–H bonds. Since the molecular formula is C_4H_8, no arene group is present. PR(p)

(ii)

G: (H)(H)C=C(H)(CH_2CH_3)

H: *Cis*: (H)(H_3C)C=C(H)(CH_3) and *Trans*: (H)(H_3C)C=C(CH_3)(H)

The *E*/*Z* isomers only gain marks if they are correctly labelled. PR(m)

(iii) But-1-ene ✓ PR(d)

(iv) The 'fingerprint' region of the spectra of the two stereoisomers will be different ✓. PR(q)

(v) Elimination ✓, because a molecule of water is lost from each alcohol molecule ✓. PR(h)